The World of science

HEAT, FUEL AND POWER

BAY BOOKS LONDON & SYDNEY

1980 Published by Bay Books
157–167 Bayswater Road, Rushcutters
Bay NSW 2011 Australia

National Library of Australia
Card Number and ISBN 0 85835 271 0
Design: Sackville Design Group
Printed by Tien Wah Press, Singapore.

HEAT AND ENERGY

Heat is essential to our lives. Without warmth we would freeze in winter and without heat energy we would not be able to use many forms of power. Heat cooks our food and without heat we would not be able to make iron and steel for such things as motor cars and refrigerators.

We can feel heat, or the absence of heat, by the effect it has on our bodies. This is a *sensation.* But in physics heat is described as the rapid movement or vibration of the molecules of which matter is composed. Heat is a form of energy in motion which is called *kinetic energy.* The hotter something is, the faster its molecules are moving. The colder it is, the slower its molecules are moving.

Measuring heat

To measure heat, we use a thermometer, which gives us a measure in Celsius or Fahrenheit degrees. A thermometer really measures the average speed of movement of the molecules in the substance. Another type of heat measurement is of the amount of heat in a substance, which means the total kinetic energy in it. This

All the thermometers below-the maximum and minimum (left), clinical (centre) and domestic (right)-are used to measure temperature. They all work on the same principle of the expansion of mercury, within a tube. As heat increases(decreases), the liquid expands(contracts) and rises(falls) to indicate a higher(lower) temperature.

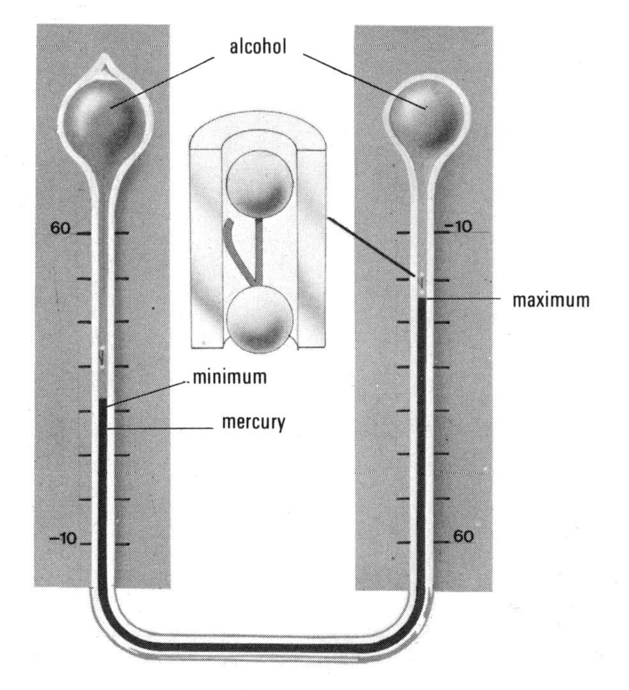

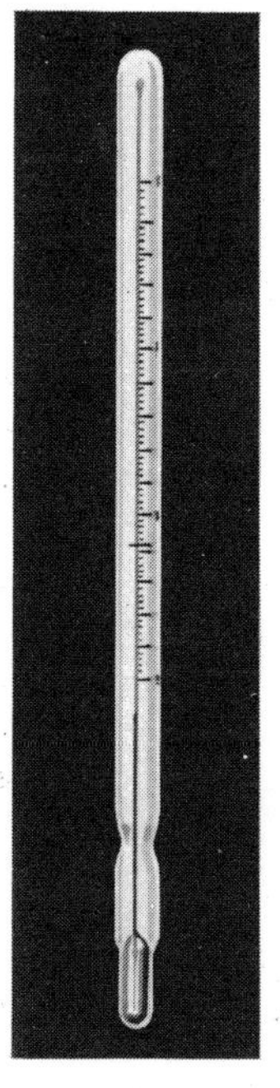

clinical thermometer

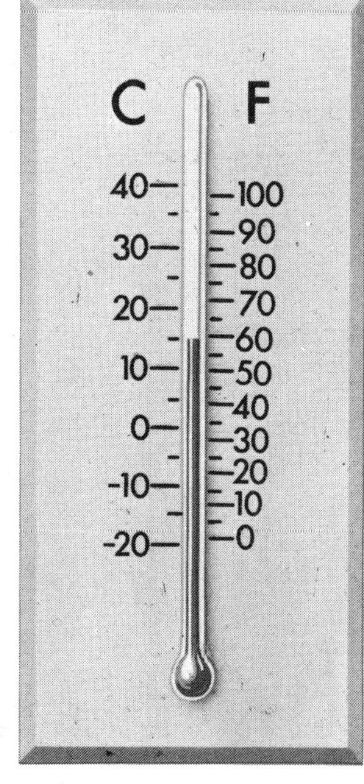

thermometer

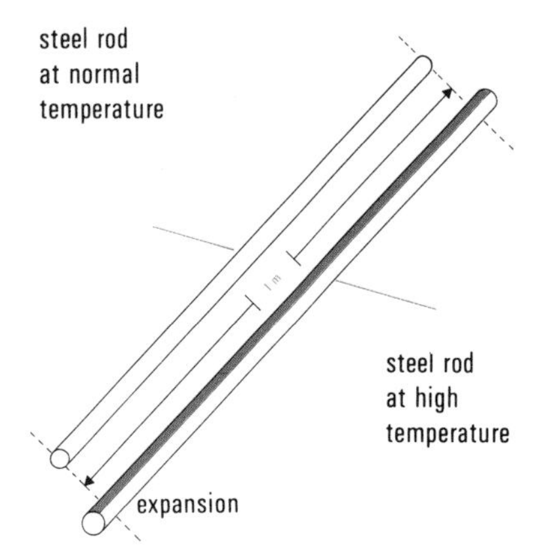

When a steel rod is heated at a high temperature, it expands.

Opposite: These sightseers are enjoying the spectacle of an enormous gushing geyser.

Below: The steelworker, in his helmet and protective clothing, is working in the intense heat of a furnace.

is the heat which the substance is capable of releasing over a period of time. The measurements used are either *calories* or *joules.*

All substances, even ice, have some heat energy, though it may be extremely small. *Absolute zero* is the lowest temperature a substance can reach. At this point (-273.16°C or 0° Kelvin) all atoms and molecules stop moving and the substance is totally without heat. In practice nothing has yet been cooled down to this point, though extremely low temperatures are reached in some research laboratories.

The molecules of hot objects give off *infra-red rays,* and objects which absorb these rays become heated. For this reason, infra-red rays are often called heat rays.

Effects of heat

A common effect of heat is that it causes substances to expand. This is true of all gases and most liquids and solids. As the substance warms up its molecules start to move faster and they need more room to move around. The amount of expansion depends on the density of the substance being heated, that is, how closely packed its

molecules were before the heating process began. Gases expand most, liquids less and solids least of all.

Heat may cause changes in the state of matter. It will turn a solid into liquid, such as ice (a solid) into water (a liquid). It will also turn a liquid into a gas or vapour, such as water (a liquid) turning into steam (a vapour). Above 6000°C all substances are gases. At temperatures of millions of degrees, such as occur in stars like the sun, substances exists as *plasmas.*

Movement of heat

Heat travels in three main ways, by *conduction, convection* and *radiation. Conduction* is the movement of heat energy from molecule to molecule. It occurs in solids and to some extent in liquids. When you put a cooking pot with a metal handle on the stove, the handle gets hot as well as the pot because the molecules of the heated pot transfer their movement to the molecules in the handle.

Convection takes place in liquids and gases. It is the movement of heated matter from one part to another. When the air in front of a fire becomes heated, it expands and becomes lighter. Colder air in the room, which is denser and heavier, will push its way under the heated air and force it upwards. Thus, currents of moving air will be formed called *convection currents* and the heat will be transferred from one part of the room to another.

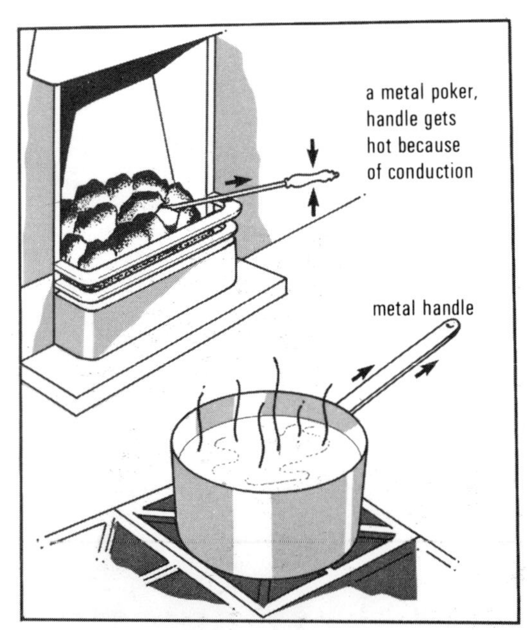

Below left: Conduction of heat. Metals conduct heat, so when you heat the end of a poker in the fire, or a pot on the stove, the heat is conducted from one part to another and even the metallic handles are hot to touch.

Below right: Radiation of heat. Heat can be transmitted through air and space as well as through solid substances. Thus, an open fire radiates heat out into a room.

Opposite centre: Convection currents take place in the atmosphere. During the day, the land is hotter than the sea and warm air rises over the land to be replaced by cooler sea air, so the breeze blows towards the land. At night, the land cools down quickly and the breeze blows out to the warmer sea.

In both conduction and convection, heat is transmitted by moving particles of matter. But heat can also travel through space, where no matter exists. It can also travel through glass, as we know when the sun shines through a window. It does this by radiation in the form of infra-red rays.

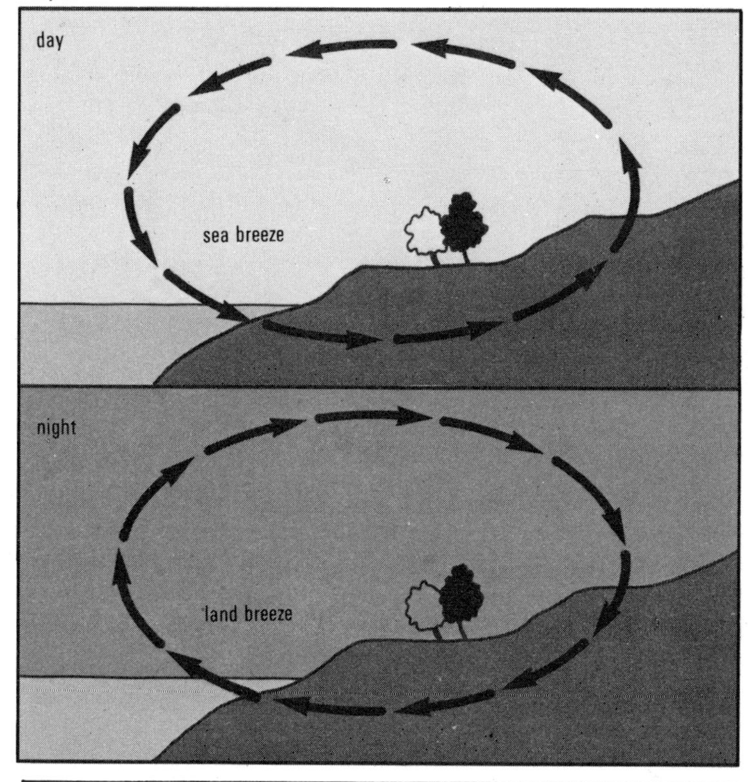

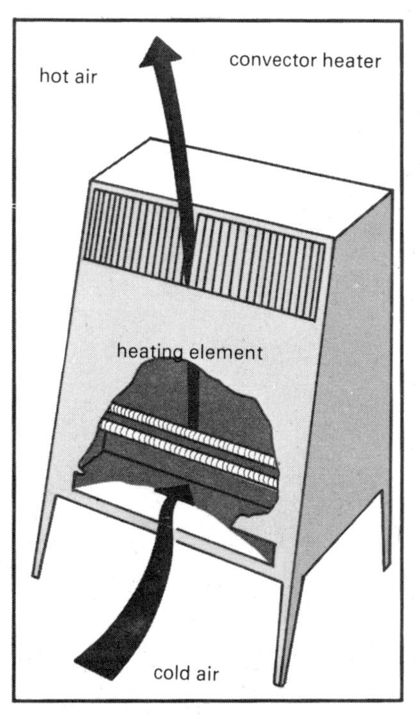

Above: Cold air enters the convector heater and expands and rises when it is heated.

Below: A flame burns strongly until covered. Burning can not occur without oxygen present.

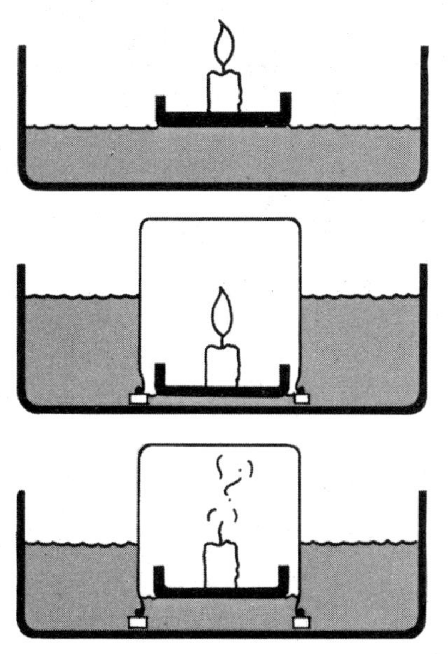

Combustion

The process of burning occurs when a substance combines quickly with oxygen in the air. This is the process of *oxidation* or *combustion.* To make a fire, there must be a good supply of oxygen, usually provided by the air, and a substance that will combine quickly with the oxygen. Such substances are *fuels.* Most fuels will not burn at normal temperatures but have to be heated to what is called *kindling point.* Some substances, such as petrol and paper, have low kindling points. Others, such as coal, need to be heated considerably before they will burn. Getting a barbecue fire going usually involves the use of fuels with a low kindling point, such as paper, wood or kerosene to ignite fuels with a higher kindling point, such as charcoal.

11

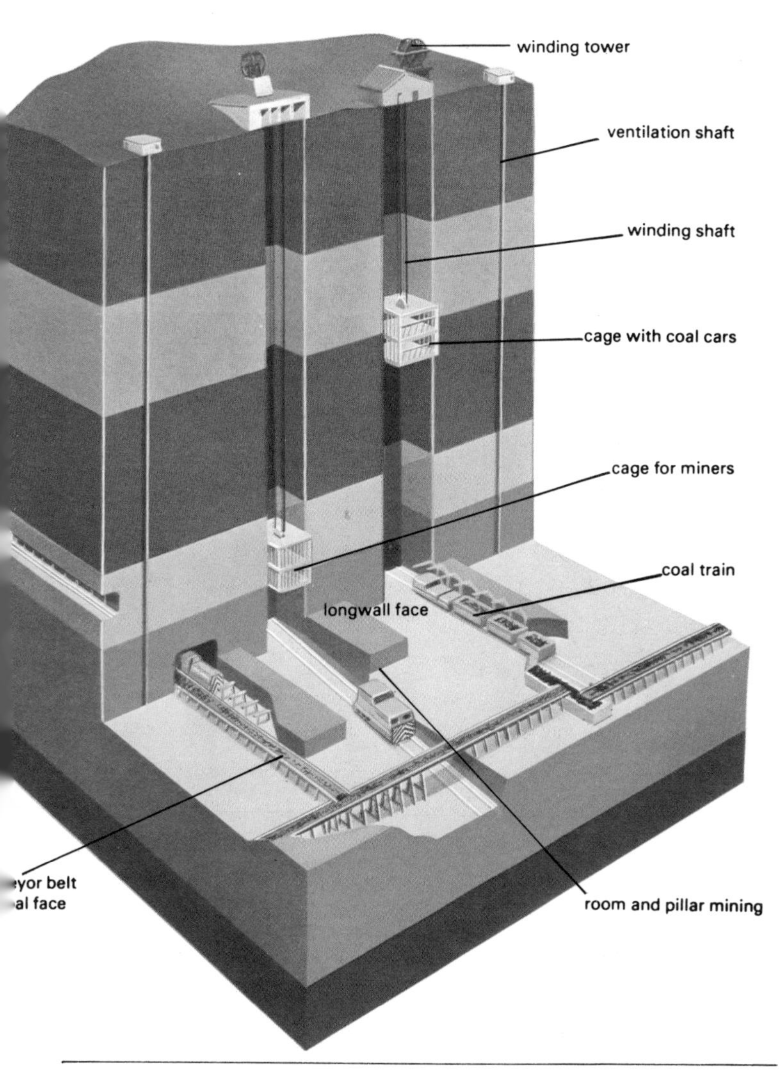

Opposite: Working in a coal mine is exhausting and dirty, even with mechanical cutters and modern hydraulic props to support the roof.

Left: A cross – section through a typical coal mine shows how air is circulated. A large fan at the surface blows fresh air down one ventilation shaft, and sucks out the stale air through another shaft.

Below: Peat, a kind of coal which is the waterlogged remains of plants, is found in bogs. It contains 60 per cent carbon(less than coal) and is dried and burnt as fuel.

Sources of heat energy

Most of the heat energy we use at the present time comes from what are known as *fossil fuels.* These are coal, petroleum or oil, and natural gas.

COAL

Coal, which is a solid fuel, has been used for providing heat from as early as 400 BC. With the Industrial Revolution in the 18th and 19th centuries, coal became the basis of manufacturing. It provided the energy to drive the steam engines that kept the factories going, turned water into steam to generate electricity and also

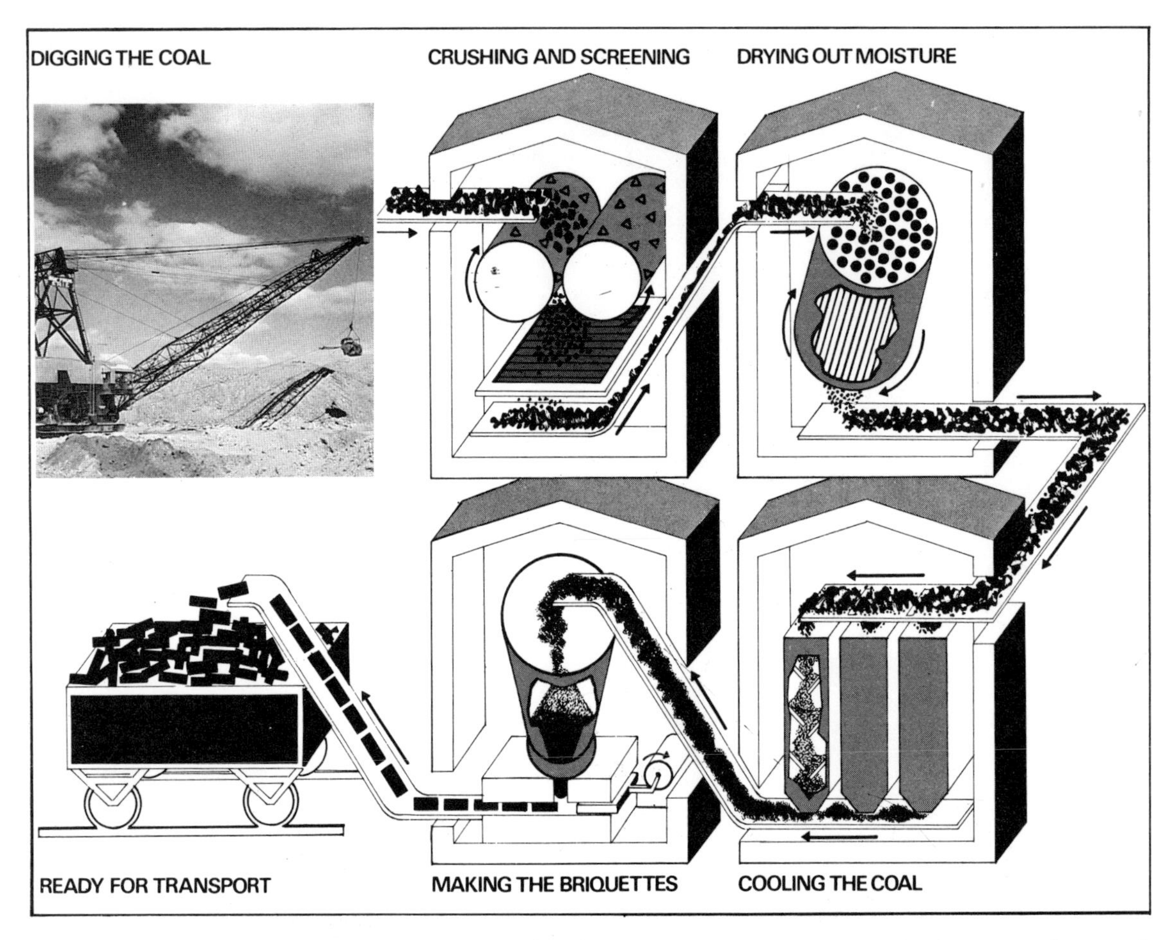

How briquettes are manufactured: the cut coal is transported on an enormous conveyor belt. It is crushed, then screened, dried and cooled. The gritty material is pressed into moulds and cut into finished briquettes.

powered ships and trains. With the last few decades coal has been replaced in many places by petroleum products and natural gas. These are cleaner to handle and give out a great deal more heat when burned. But there seems to be more coal in the ground than petroleum, so coal has once more become an important source of energy. Coal is now often processed before burning, so that it can be used more efficiently and it is even possible that ways will be found to produce petrol for motor cars from coal.

Like petroleum, coal is also a source of raw materials for products such as plastics, explosives, dyes, and synthetic rubber.

How coal was formed

Coal consists mainly of compounds of carbon, hydrogen and oxygen, water, and very small amounts of other elements like sulphur. It is *organic* in origin, that is, it was

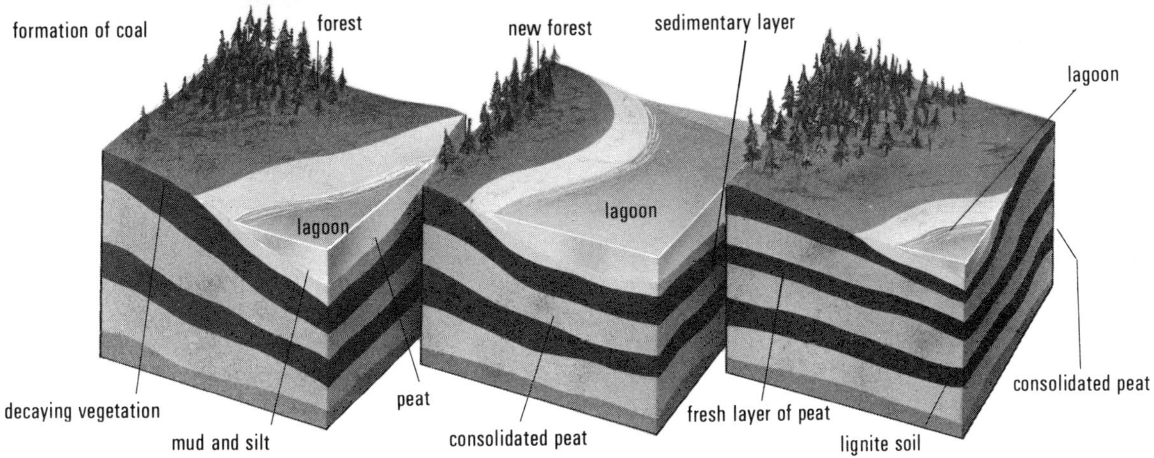

formed from living things.

Coal comes mainly from giant plants that grew on the earth about 300 million years ago during what is called the Upper Carboniferous Period. During this period, many parts of the earth were covered with shallow, swamp-like seas in which grew huge trees, giant ferns, outsize mosses and plants called horsetails.

During the Upper Carboniferous Period, the climate was warm and moist. The huge plants and trees regularly shed their leaves which fell into the swamps and began to decay through the action of bacteria. Eventually the plants themselves would die and sink into the swamps or they were covered when the land subsided as a result of movement of the earth's crust. Over a long period of time, layer upon layer of material gradually built up. Then mud and sand covered the layers of decaying vegetation, compressing them and pushing out some of the moisture.

Top: Lush forests covered the earth about 300 million years ago in the time called the Carboniferous Period. Above: Coal was formed in three stages. 1 Decaying vege–tation accumulated in the swamps and lagoons to form peat. 2 The forests were flooded and were buried in a layer of silt and mud which eventually formed deposits of coal. 3 When the floods receded, new forests grew up, more peat formed and a new cycle began.

Then, following more movements in the earth's surface, the land rose above the level of the seas and more vegetation grew in the same areas. These plants and trees also died, decayed and were covered with sediment. The cycle repeated over millions of years, with the layers increasing. As the weight of mud and sediment pushed the layers further down they become harder and the decayed and compressed vegetation became coal. Coal, then, is a fossil fuel and sometimes the outlines of leaves can be seen on a lump of coal.

Coal seams or beds

The layers of coal, called *beds* or *seams,* vary in thickness from a few centimetres up to 30 metres or more. They may lie in horizontal layers or be folded by movements of the earth's crust. Some coal seams can be seen on the surface, having become exposed through the effects of erosion where the wind and rain have carried away the covering of soil. Other seams lie deep in the ground, as far as three hundred metres down or more. Coal is mined from both shallow and deep deposits.

Kinds of coal

There are many different kinds of coal. *Peat* is a wet, fibrous material which is the first stage of coal formation.

This diagram illustrates the carbonisation process whereby coal gas is produced. When the coal is heated, it produces some gases and is converted to coke. The carbonisation of coal into coke is accompanied by the removal of tar and the gases ammonia and hydrogen sulphide. The coal gas eventually produced is used for heating.

How coal gas is produced

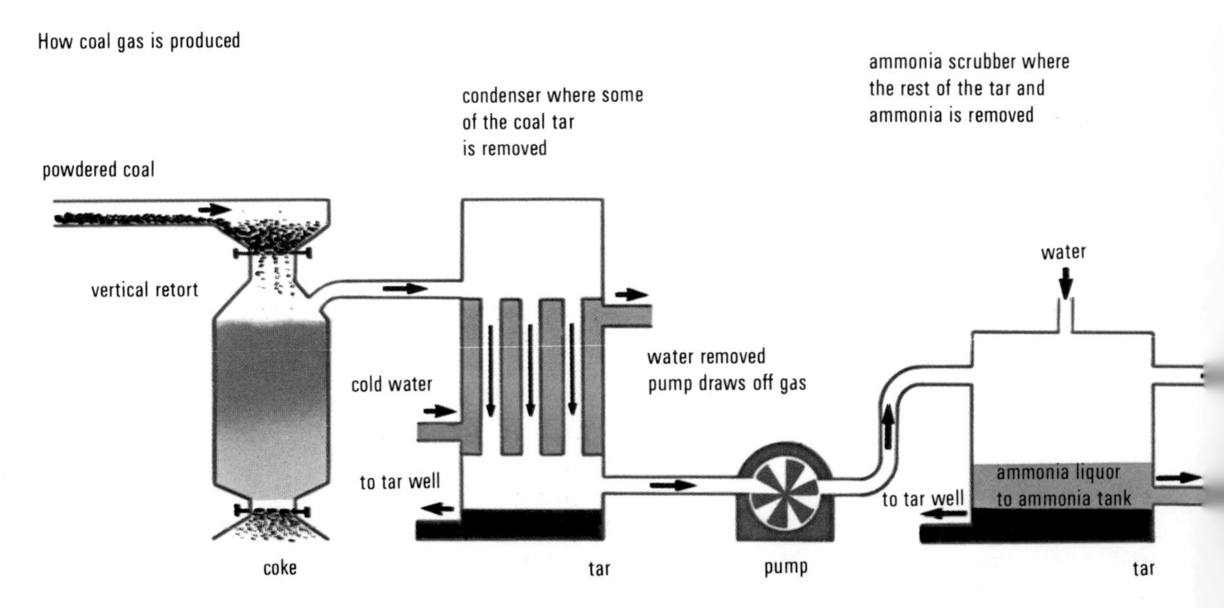

It can be found on the surface of the ground and has been used as a source of heat in Ireland, for example, for hundreds of years. When taken from the ground, peat may contain up to 90 per cent of water, which means it has to be dried out before it can be used. *Lignite* is the next stage of coal formation; sometimes it is black in colour, sometimes brown. Lignite can contain as much as 50 per cent water and must also be dried out before burning.

Black coal is often classified as *bituminous coal.* It contains far less water and much more carbon than other coals. Bituminous coals are the most plentiful, and the most used of coals. They contain little water, have a carbon content of about 90 per cent and burn well. Bituminous coals are black in colour, with dull and shiny bands. Not so common is *anthracite,* the final product of coal formation. It has practically no moisture content at all and contains about 95 per cent carbon. It is black, shiny, clean to handle, and is the best fuel of all coals. It also causes least pollution when it is burnt.

Mining coal

Coal seams on or near the surface are *strip mined,* or mined by *open cut methods.* Bulldozers are used to strip off any soil covering, called the *overburden.* Then the coal is broken into manageable pieces, sometimes by explosives. The coal is then loaded onto trucks for

Lignite

Bituminous coal

Anthracite

Above: Different types of coal. Lignite, which may be brown or black, is mined by open cut methods; bituminous coal burns well and is used for heating: anthracite is the top-quailty form of coal.

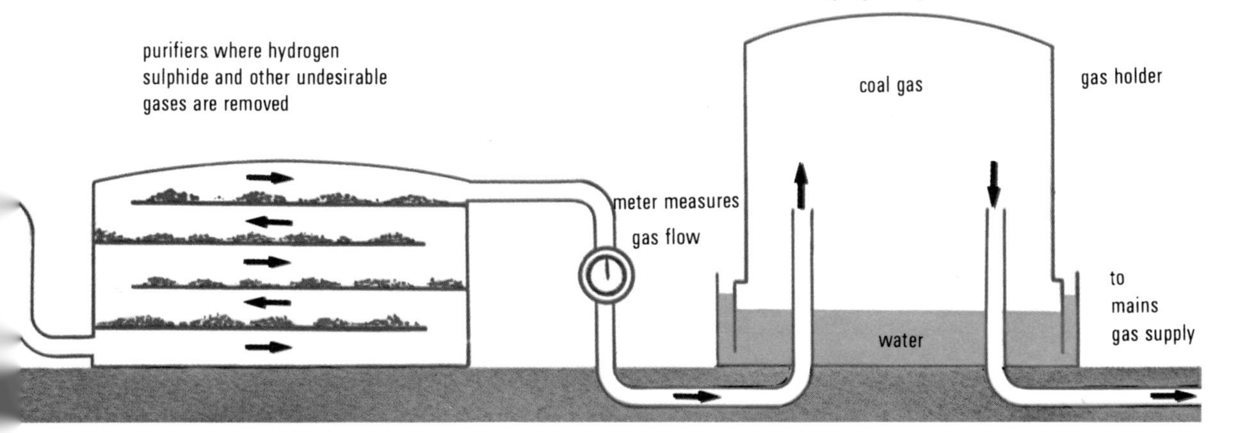

Above: This underground shuttle car tansports the cut coal to the trucks which carry it upwards to the coal face.

Above right: This weary miner's face is coated with grime and coal dust after a hard day's work underground in the mine.

transport, often to thermal power stations, where it is used to generate electricity.

For deeper seams, shaft mining is used. Shafts are sunk into the ground and when the seam is struck, miners extract the coal by following the direction of the seam. A common method of underground mining is called *undercutting.* Miners make a cut at the bottom of the coal face and then drill holes over the face, insert explosive charges and detonate them. This is called shot-firing. The coal face then crumbles and the pieces of coal are taken away on conveyor belts.

Nowadays, much coal is mined with mechanical coal cutting machines which take a great deal of the labour and some of the danger out of the work.

Coal gas and coke

Obtained from coal by a process called *carbonisation,* coal gas is used as domestic fuel in gas stoves and industrial heating. Coal is heated in large ovens called *retorts* but is prevented from burning by shutting out the air. At a temperature of 1200-1300°C the coal gives off a mixture of gases and vapours. The gases and vapours are treated by 'washing' to remove tar and other substances including ammonia and hydrogen sulphide. These products are valuable by-products of the carbonisation process. The gas is stored in the big tanks we call gas tanks, or gasometers. The whole process takes about 12 hours.

What is left of the coal after the gas-making process is

coke. It is grey and porous and contains 90 per cent or more of carbon. Coke is used as a fuel in industry, where it is essential for certain processes including the smelting of some ores to form metals.

PETROLEUM

Petroleum or mineral oil is one of the most valuable substances found in the earth's crust. It is a mixture of a number of different hydrocarbons from which gasoline, fuel oil and lubricating oils are produced. Petroleum is also used to manufacture a wide range of products including plastics, paints, drugs, explosives, cleaning fluids and detergents.

How petroleum was formed

Petroleum was probably formed from algae and other small plants and animals that flourished in the seas millions of years ago. Their remains were buried in clays brought down by rivers and were decomposed into simple *hydrocarbons* by the action of bacteria. Hydrocarbons are compounds of hydrogen and carbon. Eventually the clays were buried under *permeable* or

This simple diagram explains how coal and oil were formed. The chemical energy these fuels contain today was solar energy 300 million years ago when the cycle began.

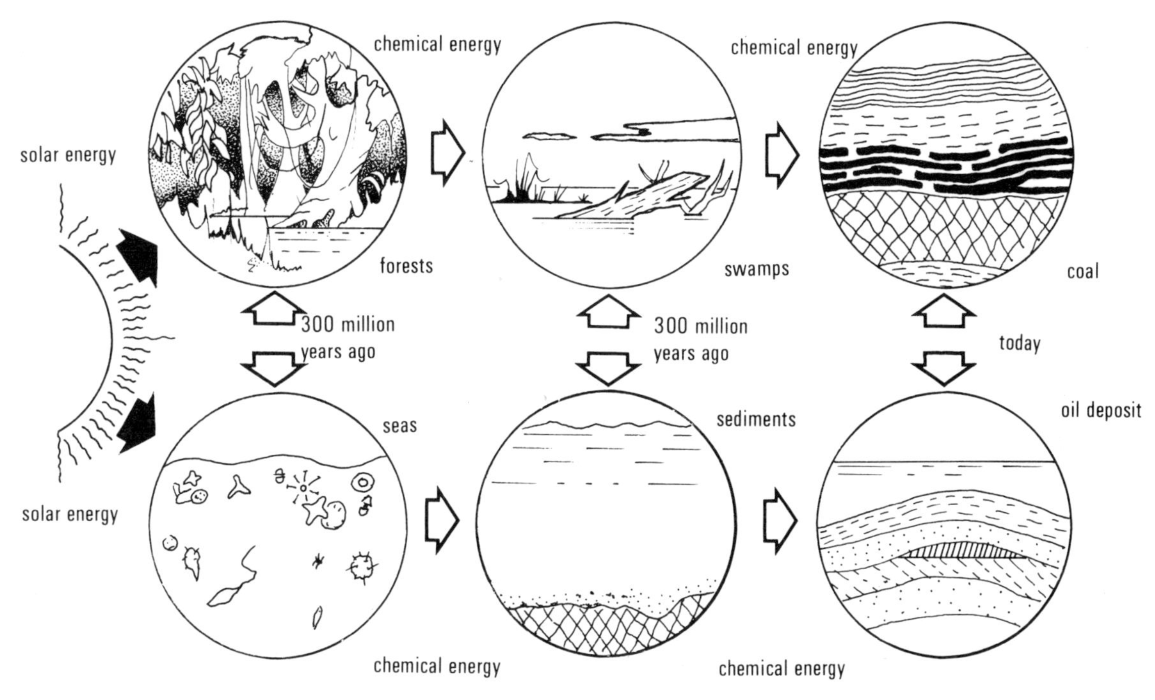

porous rocks and were compacted by pressure, causing carbon compounds to change into hydrocarbon oils. These oils flowed in the permeable rocks and formed giant pools or underground oil reservoirs. Not all petroleum is exactly the same and the differences may be the result of different temperatures and pressures at the time the particular deposit was formed.

The usefulness of petroleum has been known for quite a long time. Oil from the ground was burnt to produce light and used for treating roads, for waterproofing and in medicine. At first, people used the oils that seeped out of the ground, then they drilled holes by hand. The first mechanical drilling for oil was done by Edwin Drake in Pennsylvania, USA in 1859.

Above: This is not a handwoven tapestry but a visual interpretation of layers of petroleum-bearing rock. The white area is a salt dome. Below: This is a photograph, not an impressionist painting, of a de-gassing station. The burning gas flares have created a heat haze.

Where petroleum is found

The underground reservoirs containing giant pools of oil are usually found where there is an *anticline* in the earth's crust. This is an upfold of rocks which forms a kind of dome and acts like a cap sealing the oil inside the reservoir. Though petroleum deposits form in permeable rocks, the rock around the dome must be *impermeable,*

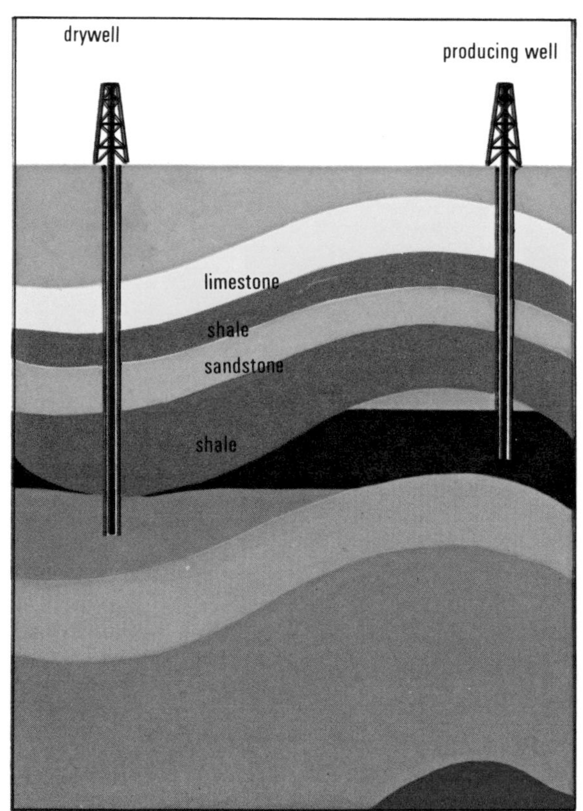

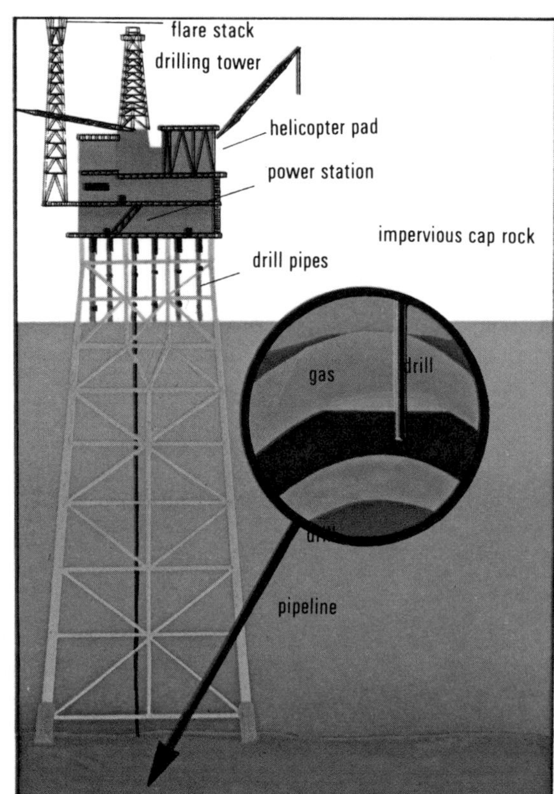

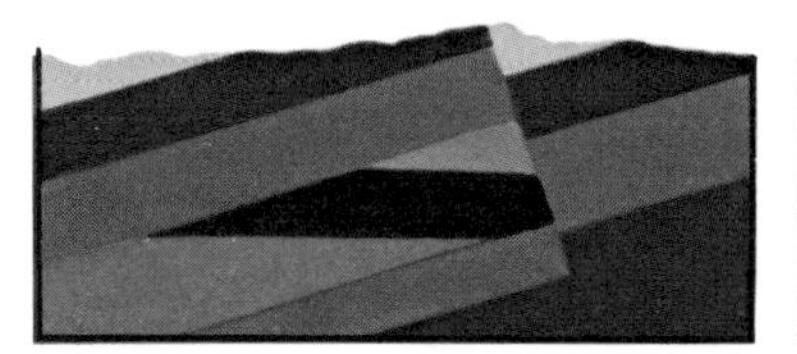
fault trap

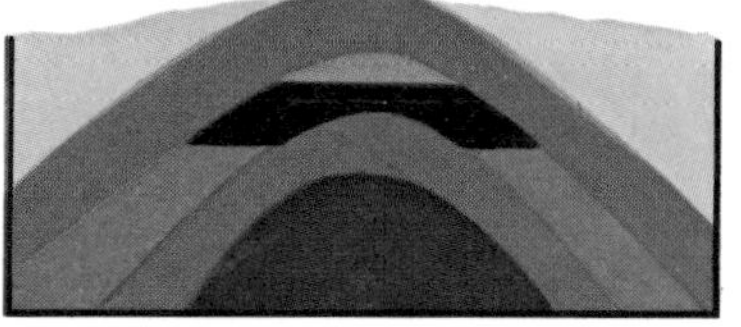
anticline trap

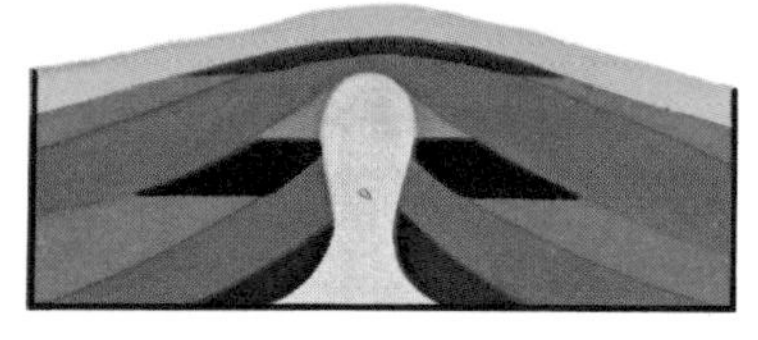
saltdome

otherwise the oil would flow away.

Places in the earth's crust likely to contain reservoirs of oil are found by geologists and geophysicists. From a study of the rocks on the surface geologists are often able to make a guess at what may lie underneath. Geophysicists often set off small explosions and measure the shock waves that pass through the rocks under the ground. Different types of rocks reflect the shock waves in different ways and so it is possible to work out the type and structure of the rocks in the area.

Using these and other methods, scientists can tell if there is the possibility that a reservoir of oil might exist. But the only way to be really sure if there is oil in a particular spot is to drill a hole.

Top left: Two oil wells can drill in the same field but only one may strike oil.
Top right: This diagram shows an oil-production platform with(inset) a cross–section view right through the sea bed.

Above: Fault traps, anticline traps and salt domes in the earth's crust may contain oil if it lies on impervious rock.

Drilling for oil

Below: These two men working on an oil platform are drilling.

Below right: Oil platforms are towed out to sea and secured to the sea bed. This oil tanker is being filled at sea.

Most drilling is now done with a rotating shaft held in a tower called a *derrick.* On the end of the shaft is a bit which has a series of sharp metal teeth. Different kinds of bits are used, depending on the type of rock to be drilled. The bits crush the rock to fragments which are then washed out with a mixture of water and clay called *bentonite.* The bentonite also seals the walls of the hole and keeps the drilling bit cool. When geologists want to study the rocks being drilled, a *coring bit* is used. Coring bits have diamond cutters around the edges to cut through the rock and a hollow centre which holds a sample of the rock, the *core.* When the bit is raised to the surface, the core can be examined.

Usually, petroleum is found floating on salty water and often a layer of natural gas is found on top of the petroleum. If the reservoir is under great pressure as it often is, the gas may be dissolved with the petroleum and may come out with it. Sometimes the pressure will force the petroleum out in a *gusher.* Gushers, however, are now usually controlled as much petroleum is wasted and it causes pollution. When the pressure is no longer sufficient to force the petroleum up the pipe, pumps are used. The final stage is to flush the petroleum out with the water. Even so, much of the petroleum, possibly as much as half, cannot be recovered.

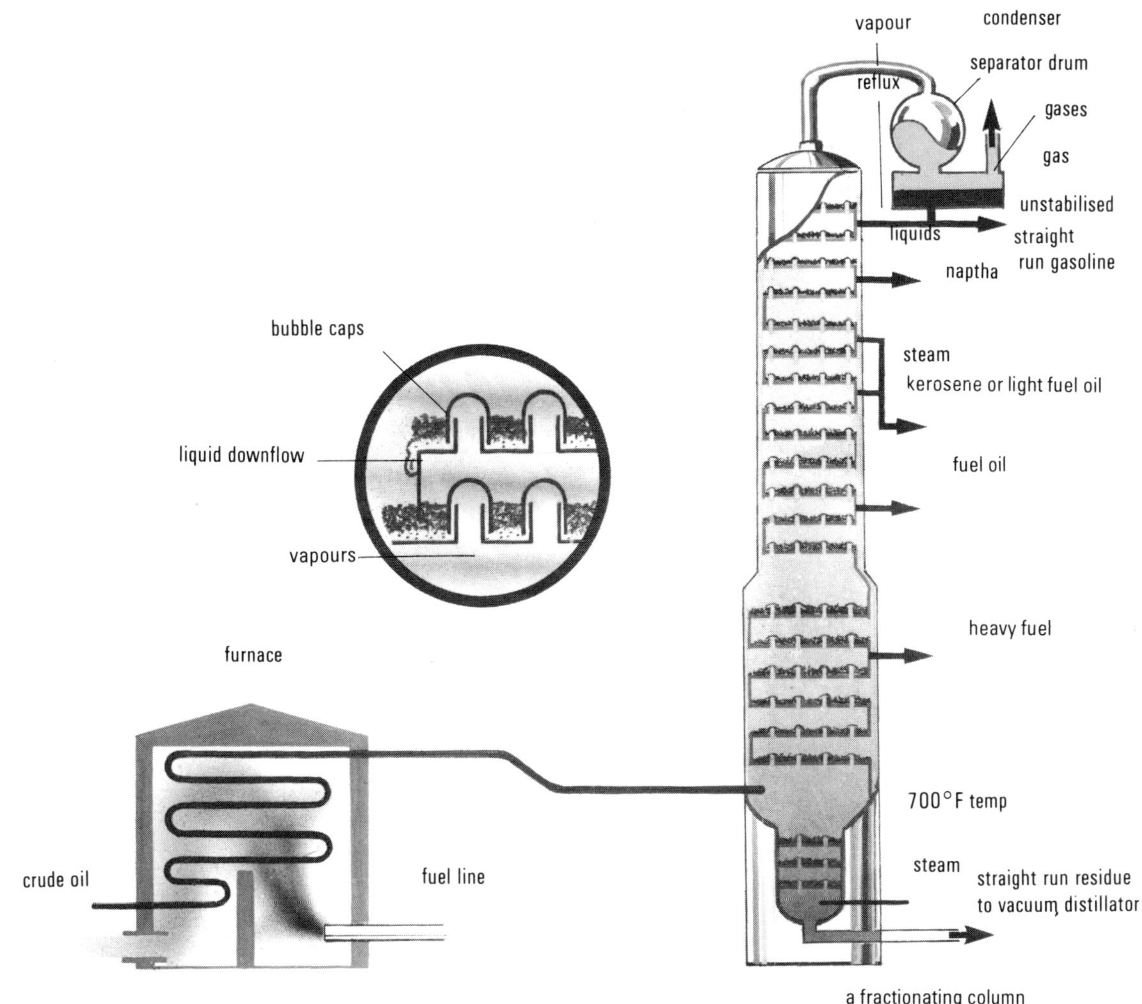

a fractionating column
used for petroleum distillation

Refining petroleum

'Crude' petroleum, which is oil as it comes from the ground, contains many different compounds. To make petrol, kerosene, heating and lubricating oils, petroleum is first heated in a furnace and the gases are passed into a tower called a *fractionating column.* The column is hottest at the bottom, getting cooler towards the top. The boiling point of the various compounds in the petroleum varies. It depends mainly on the weight of their molecules, which is largely the result of a number of carbon atoms in them. When petroleum boils, the vapours travel up the

This fractionating column is used to distribute petroleum. The petroleum boils at the hot base of the tower, and the vapours travel up to the cooler top before condensing into liquids.

Above: The big yellow sheet of flame is burning from wastage gas at the surface.

Right: Natural gas is relayed across land to domestic and industrial consumers by a vast network of gas pipes.

fractionating column before condensing again into liquids. The different substances, or *fractions,* condense at different levels and are collected in containers called *bubble cap trays,* which are placed at intervals up the column.

The lightest fractions supply gasoline or petrol. Increasingly heavy fractions give kerosene and fuel oil. Fuel oil cannot be further separated by boiling in air, so it is reheated in a vacuum where the fractions boil at a much lower temperature. This reboiling provides diesel oil, lubricating oil, asphalt and paraffin wax.

Making petrol and diesel oil

The main use of petroleum today is as gasoline or diesel oil for engines. But crude oil does not yield enough of these lighter fractions. To increase the supply, the heavier fractions are broken down into lighter ones by *cracking.* Cracking is done by heating the heavier fractions under pressure, sometimes using a *catalyst,* which is a substance which helps a chemical change in another substance, but does not change itself. In this process some *alkenes* are formed which improve the quality of the gasoline.

Crude oil contains many useful substances and many more can be obtained when it is further refined. Some of these compounds are the tarting materials for the *petrochemical industry* which produces plastics, fabrics, drugs and other products. The various processes of distilling, or fractionating, vacuum distilling and cracking can be altered to produce differing quantities of the various compounds and the fractions can then be mixed to give products with special characteristics.

Natural gas

One of the world's most important sources of energy, natural gas is usually found with petroleum, having been formed in a similar way millions of year ago.

Natural gas heats about twice as efficiently as manufactured coal gas and, unlike coal gas, natural gas contains no carbon monoxide or other poisonous gases. It also has no odour or smell so small amounts of strong-smelling substances such as organic sulphur compounds are added so we can detect gas leaks before dangerous leaks lead to explosions. The different burning characteristics of natural gas compared with manufactured coal gas mean that different burners have to be used in stoves. Therefore, when natural gas is first introduced into a city all the stoves have to be converted.

Enormous oil tankers, built to meet the increased demand for crude oil, transport oil around the world. If a tanker should sink and spill its cargo, there may be considerable pollution with marine life and many seabirds endangered.

Natural gas can be purified to produce two other types of gas, *propane* and *butane.* Both these can be made into liquids under pressure, when they are called *liquified petroleum gas* (LPG), or bottled gas. Anyone who has been camping or spent a holiday in a remote week-ender will know how convenient bottled gas can be for cooking, lighting and running a refrigerator.

LPG is now being used as a substitute for petrol in some cars and trucks. Although the fuel supply system has to be modified to allow the engine to burn LPG, there are savings in running costs which can soon make up for the cost of conversion.

Pipelines

Both petroleum and natural gas are often carried in pipes from the source of supply to some other place. In the Middle East, for example, some of the oil fields are inland and the oil is carried by pipes to the ports where it is pumped aboard oil tankers. In Australia, natural gas is carried from natural gas fields along pipes to the cities. One well known gas pipeline is the one that brings South Australian gas to Sydney.

Pipe-laying is now highly mechanised. A trencher or excavator first digs out the trench, throwing the soil to

How a four-stroke petrol engine works. 1 Intake-as the piston descends, petrol and air are drawn in. 2 Compression -the piston ascends and compresses the mixture. 3 Power-a spark explodes the fuel and the piston descends. 4 Exhaust-the piston ascends and drives out exhaust gas.

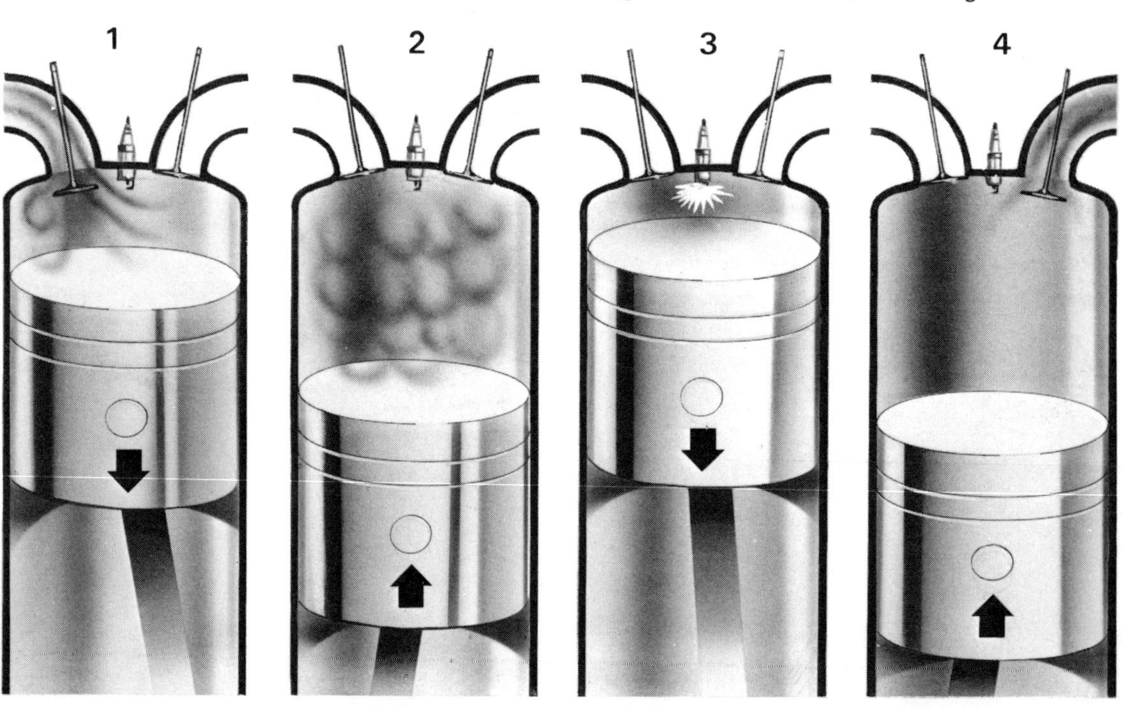

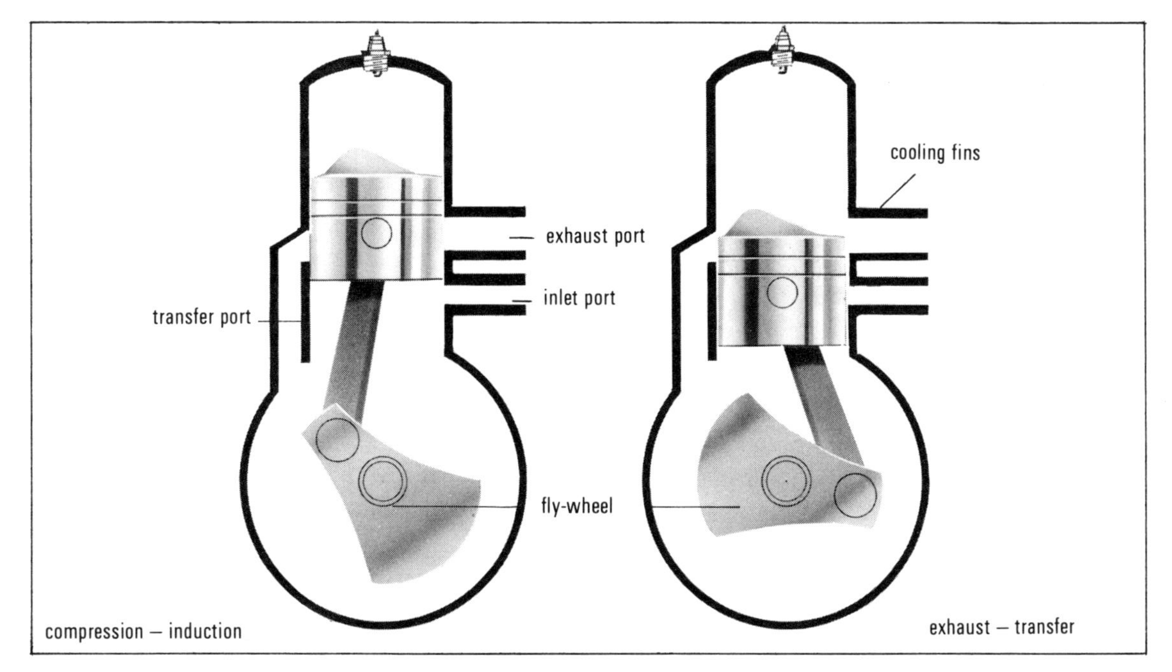

one side. Behind the trencher, welders join lengths of steel pipe together. A machine then cleans the pipe and coats it with bitumen or other waterproof material to prevent corrosion. Crawler tractors with hoisting equipment then lift the pipe into the trench and bull — dozers fill in the trench.

In any large pipeline, the fluid being transported must be pumped through the pipes. If the pipe cannot be laid completely level there will be differences in pressure along its length and at points where the pressure builds up, the oil or gas will slow down. Losses in speed also occur as the result of friction between the fluid and the pipe. Pumping stations are situated at intervals along the pipeline to boost the flow.

How a two-stroke petrol engine works. 1 Both fuel and air are drawn into the crankcase and the mixture in the cylinder is then compressed. 2 The spark explodes the mixture and, as the new mixture enters the cylinder, exhaust gases are pushed out.

THE INTERNAL COMBUSTION ENGINE

The engines of most motor vehicles, such as motor cars and motor bikes, use gasoline, or petrol, as their fuel. A petrol engine works on the basic principle that heat causes a gas or vapour to expand. It is the expansion of the heated gas that provides the energy to drive the motor.

The typical petrol engine is called an *internal combustion engine* because fuel is burned inside enclosed cylinders. A motor car engine is also called a *reciprocating* engine because it has pistons which move up and down inside the cylinders and up-and-down movement is reciprocating movement.

In petrol engines, a repeating cycle of events takes place. In motor car engines this is always a *four — stroke cycle.* In small motor cycle and outboard engines it is often a *two-stroke cycle.*

The four-stroke cycles

This diagram shows the four strokes which correspond to the stages of induction, compression, firing and exhaust in the cylinders of the four-stroke petrol engine.

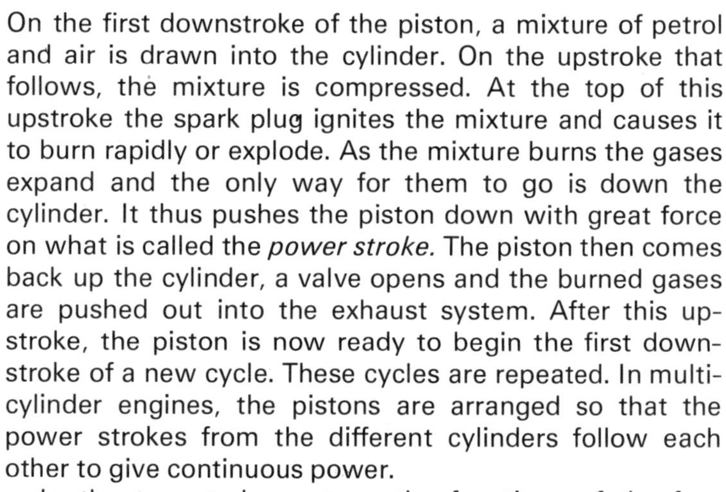

On the first downstroke of the piston, a mixture of petrol and air is drawn into the cylinder. On the upstroke that follows, the mixture is compressed. At the top of this upstroke the spark plug ignites the mixture and causes it to burn rapidly or explode. As the mixture burns the gases expand and the only way for them to go is down the cylinder. It thus pushes the piston down with great force on what is called the *power stroke.* The piston then comes back up the cylinder, a valve opens and the burned gases are pushed out into the exhaust system. After this upstroke, the piston is now ready to begin the first downstroke of a new cycle. These cycles are repeated. In multi-cylinder engines, the pistons are arranged so that the power strokes from the different cylinders follow each other to give continuous power.

In the two-stroke system, the functions of the four strokes are combined into two.

A German engineer, Nikolaus August Otto, developed a successful engine working on these principles in 1876, using gas as the energy source. The first successful gasoline-burning engines were developed by the automobile pioneers Gottlieb Daimler and Karl Benz.

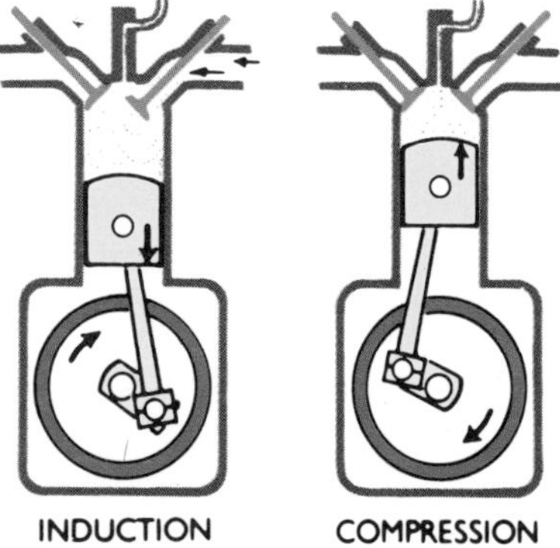

INDUCTION OF AIR

COMPRESSION

The diesel engine

Also in the 1890s, the German inventor, Otto Diesel, developed the first diesel engines, which are used in ships, locomotives, trucks and tractors, and some motor-cars. The Mercedes-Benz, for example, sometimes uses a diesel engine.

Although the basic principles of the diesel engine are the same as those of the petrol engine, there are important differences. The diesel burns fuel oil, commonly called 'distillate' which is composed of cheaper fractions than petrol, and it is particularly suitable for giving high power at low speeds.

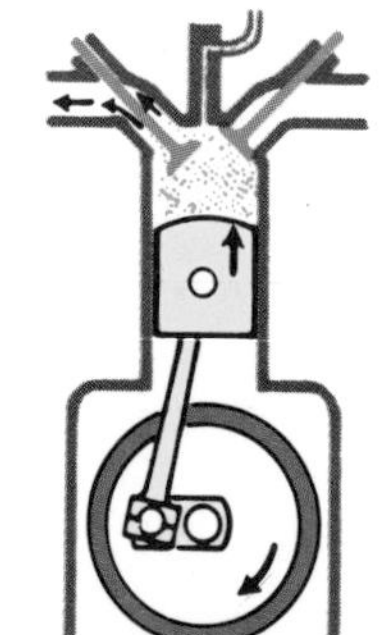

OIL INJECTION AND FIRING

EXHAUST

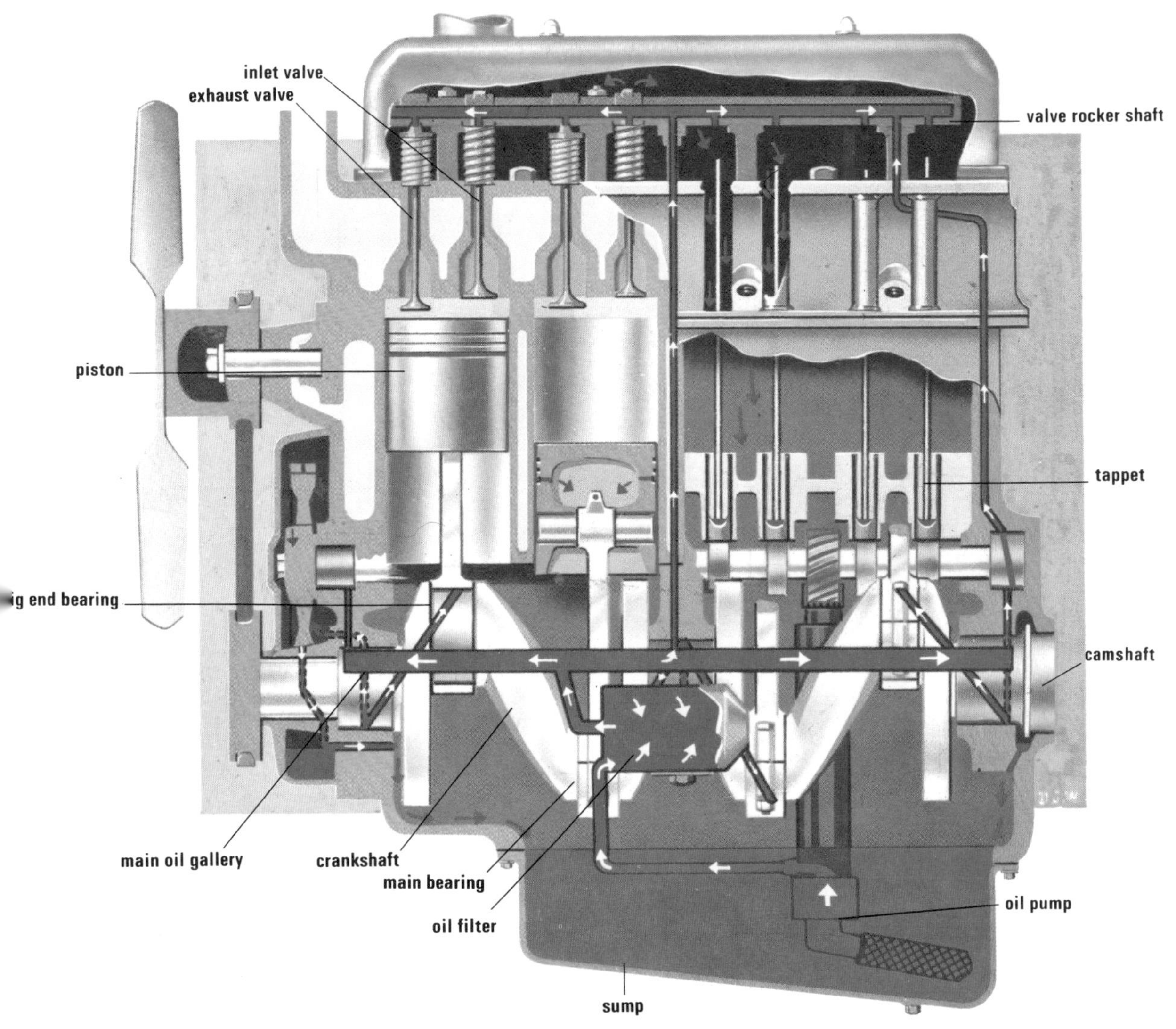

Above: This cross-section view through a conventional diesel engine clearly shows the different parts. Arrows show the direction of oil flow.

Left: This man works on an assembly line in a big factory, assembling car engines.

This large Rolls Royce engine has six cylinders which make it extremely powerful.

In the diesel engine, no spark plug is used. The fuel is injected into the cylinder at the point where the upstroke has so compressed the air that it is hot enough to ignite the fuel.

How engines work

The reciprocating movement of the internal combustion engine must be converted to a turning, or *rotary* motion. This is done by linking the pistons to a *crankshaft.* The crankshaft transmits power from the engine to the wheels through the car's transmission system, which includes the gearbox. The gearbox makes it possible to use different combinations of cogs so that the car will move fast or slowly but with enough power at all times.

A petrol engine requires a number of systems to supply its basic needs. The fuel has to be supplied in just the right

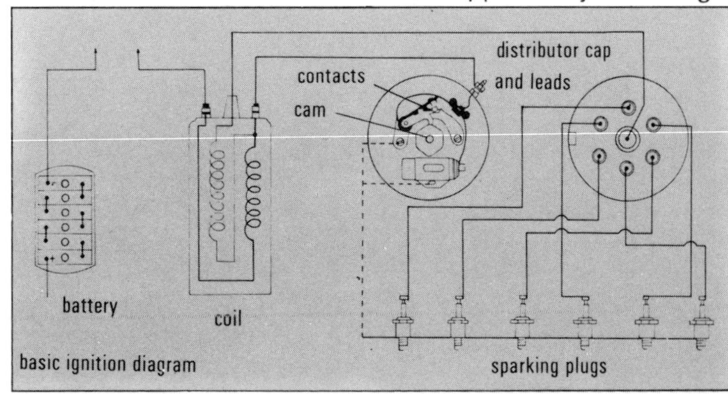

This diagram illustrates the basic ignition system in a car. The battery voltage is fed into the coil and changed into a high voltage to produce a spark which will then explode the fuel.

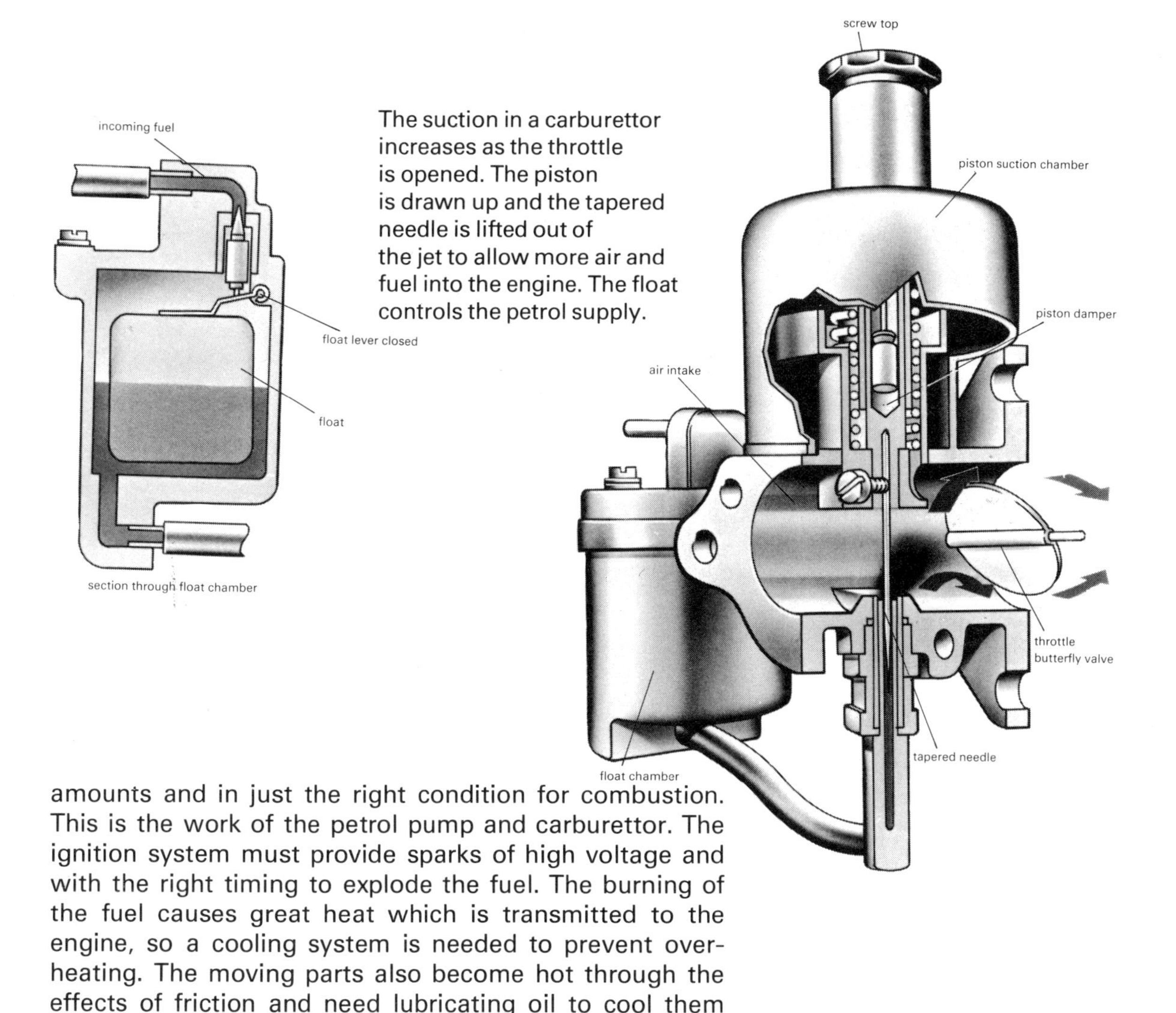

The suction in a carburettor increases as the throttle is opened. The piston is drawn up and the tapered needle is lifted out of the jet to allow more air and fuel into the engine. The float controls the petrol supply.

amounts and in just the right condition for combustion. This is the work of the petrol pump and carburettor. The ignition system must provide sparks of high voltage and with the right timing to explode the fuel. The burning of the fuel causes great heat which is transmitted to the engine, so a cooling system is needed to prevent over-heating. The moving parts also become hot through the effects of friction and need lubricating oil to cool them down and prevent wear.

Parts of an engine

The usual motor car engine is made of two basic units, the cylinder head and the block. The cylinder block is machined from a solid cast metal which is usually cast iron but sometimes is aluminium. It contains the engine cylinders which have to be very accurately finished. The pistons which travel up and down inside the cylinders are usually made of aluminium alloy, which is lightweight and tough. To prevent the gases of the burning fuel from leaking around the pistons, they are fitted with springy bands of metal called piston rings.

Engines may have any number of cylinders which can be arranged in a line, or in opposite pairs. They can form a

This gigantic engine resembles a space rocket but belongs, in fact, to a Boeing 747 aeroplane, known as a Jumbo Jet.

'V' or be opposed, that is, lying on their sides opposite each other. Many piston-driven aeroplanes have their pistons arranged in a circle.

The lower part of the cylinder block is known as the crankcase. It holds the *crankshaft* that converts the up and down motion of the pistons into rotary motion. The pistons are connected to the crankshaft by *connecting rods.* The upper, or smaller, end of the connecting rod is connected to the bottom of the piston and the lower half, or *big end,* connects with the crankshaft.

Each piston drives the crankshaft only on the power-stroke of its cycle. On the other strokes, the movement of the crankshaft is driving the pistons, but the cylinders fire in a cycle that is known as the *firing sequence,* so that there is continuous power.

The fuel system

The fuel system has to supply the right mixture to the cylinders, at the right time. The system includes a pump to transfer the petrol from the tank, which, for safety reasons, is located away from the engine. The fuel pump sends the fuel into the carburettor, which mixes the petrol with an exact amount of air. This is done through jets; the petrol is sprayed into a swiftly moving air-stream and mixes with the air.

The mixture of petrol and air is sucked from the carburettor into the *inlet manifold,* which delivers it through inlet valves to the *combustion chambers,* that is, the tops of the cylinders. This happens on the first downstroke.

The ignition system

All internal combustion engines use an ignition system to cause the fuel to burn rapidly or explode. 'Ignition' means causing something to burn, and ignition systems use an electrical spark to ignite the fuel. In a motor car, the ignition system usually consists of a battery and a coil, a generator and a distributor and a spark plug in each cylinder. When starting the car, the turning of the key in the ignition switch allows the battery to release current to the coil. The coil acts like a transformer to increase the six or twelve volts of the normal system to a voltage high enough to 'jump' the gap between the points of the spark plugs in the cylinders and fire or ignite the explosive fuel

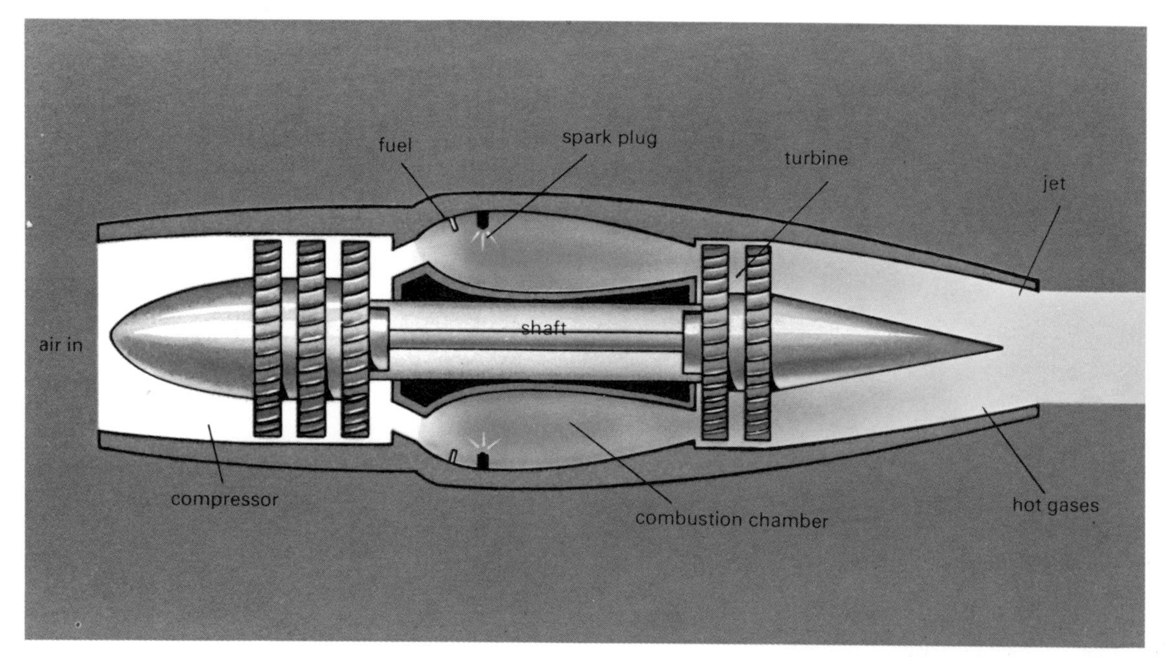

and air mixture.

The current from the coil passes first to the distributor, where it is directed to the spark plugs in the proper firing order, so as to explode the fuel in each cylinder at the right time. When the electric charge reaches the spark plug, the cylinder is filled with the compressed fuel and air mixture above the piston. This explodes, pushing the piston down, and the engine starts.

Jet engines

Jet engines also use the basic principle of heat causing a gas or vapour to expand. In a jet engine, air is compressed and forced into a combustion chamber. Here fuel is injected and ignited. The mixture of compressed air and fuel is rapidly heated by the burning process. The gases expand and can only leave the engine one way, out the back. The backward force of the gases leaving the engine at high speed pushes it forward.

Jet engines and space rockets work on the same basic principle, but there is one important difference between them. The jet engine uses the oxygen in the air to burn its fuel, just as a motor car does. But oxygen exists only in the atmosphere, and not in space. The rocket has to carry its own oxygen supply which it does in the form of liquid oxygen. This is one reason why space rockets are so huge, but it is also the reason why they can work in space.

Top: In the turbojet, fuel is burned rapidly in the combustion chamber. This causes a blast out of the back of the jet. The plane is thrust forward by the blast's reaction. Above: Aeroplanes have to be checked and serviced on a regular basis by engineers. These men are working on a huge 747 Jumbo jet engine.

STEAM ENGINES

The first practical steam engine was invented by Thomas Savery in 1698. It had a boiler from which steam passed to a cylinder, after filling with steam, the cylinder was cooled from outside with water. The steam condensed, and as it did so its volume decreased. This created a vacuum which caused water to be sucked up through pipes and these engines were used as pumps.

Opposite: In a triple expansion steam engine, which was in use until the 1940s, the steam is passed through 3 cylinders which are connected by slide valves. The pistons are linked by rods which transmit the drive to the crank-shaft to provide the driving power for marine propellers.

Below: Steam engines now exist only for holiday-makers and railway enthusiasts, but they once pulled all the trains.

The development of the steam engine

In 1705, Thomas Newcomen developed an improved engine in which the steam was cooled by a jet of water sprayed into the cylinder. The cylinder had a piston that was pushed up by the expanding steam, and pulled down by the vacuum created when the steam was cooled. In this engine, the piston was joined to a beam that was

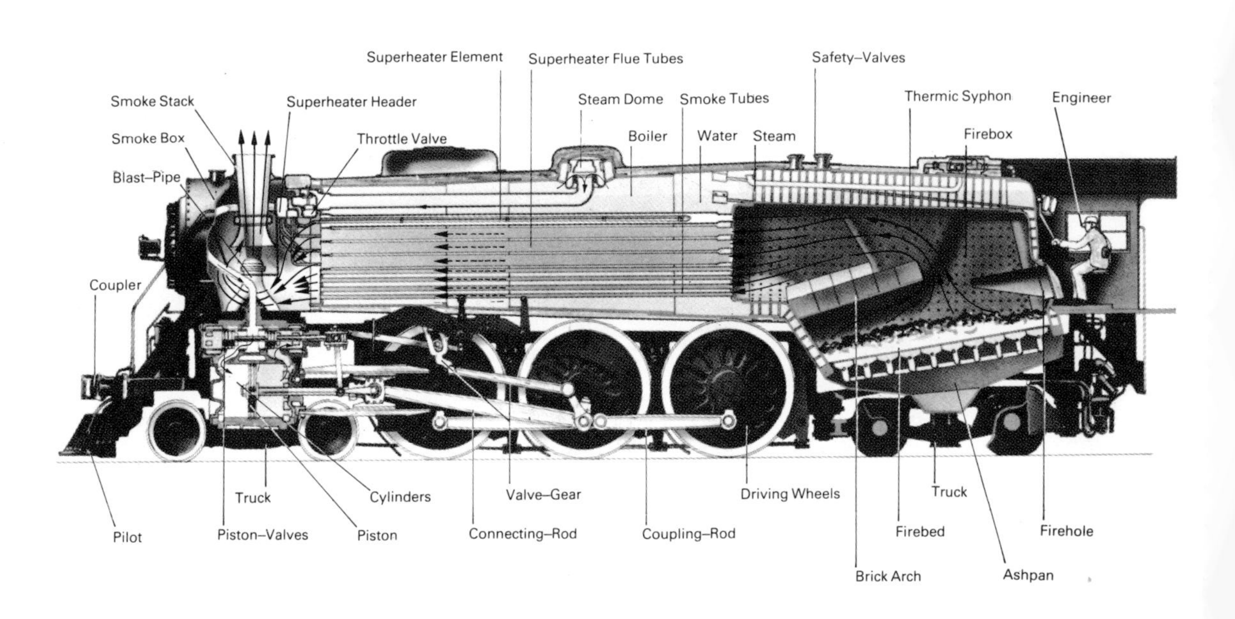

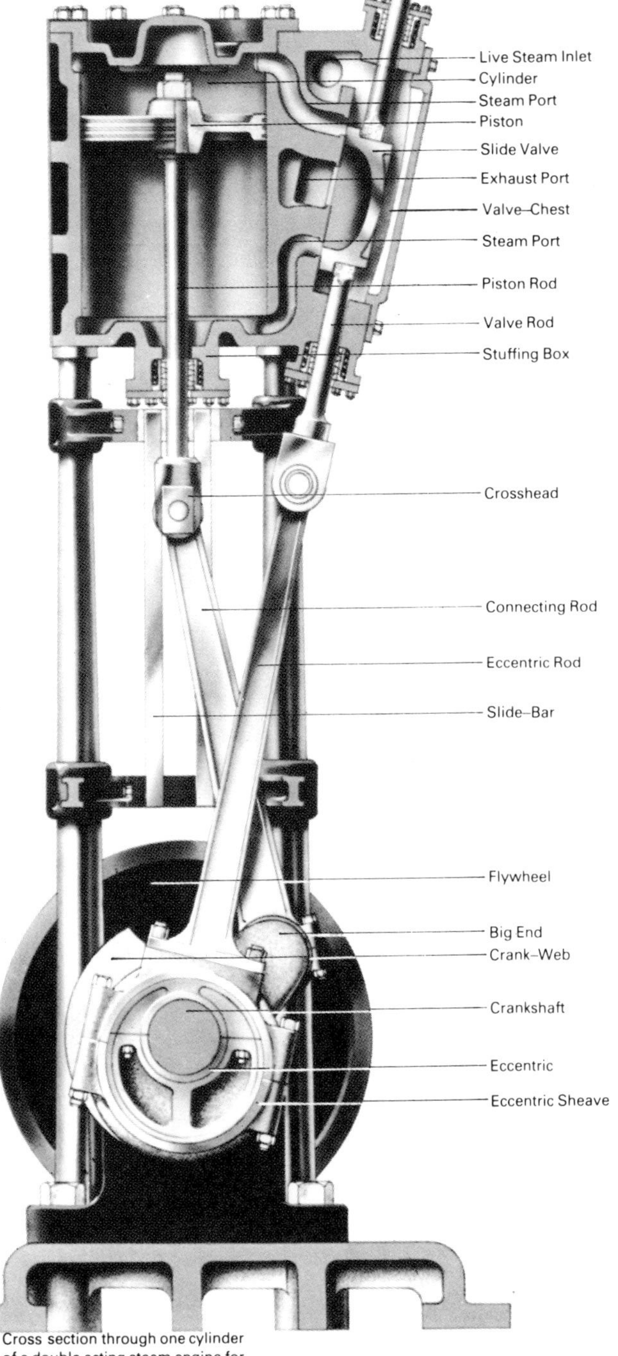

Cross section through one cylinder of a double acting steam engine for single–direction running without variable cut–off.

Low Pressure Steam   High Pressure Steam   Steam under Compression

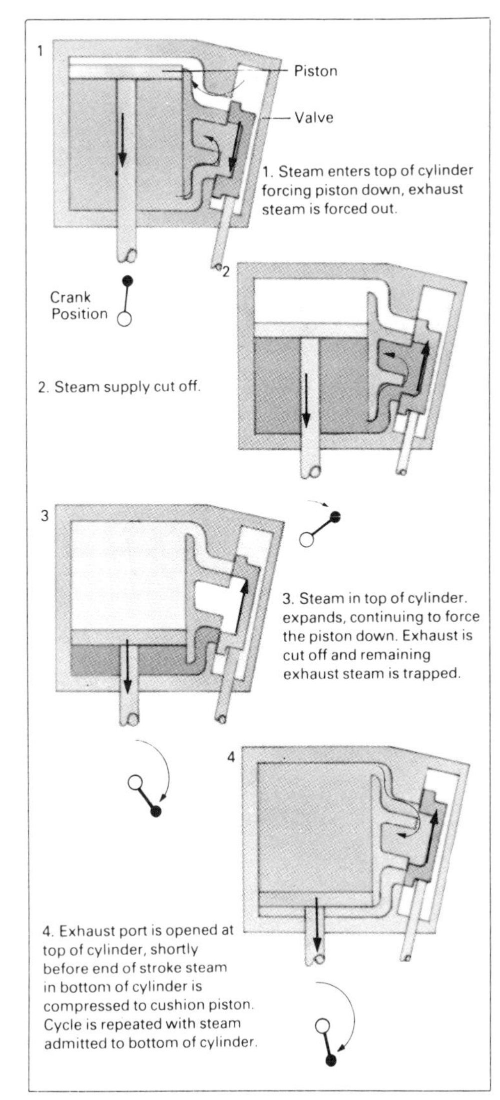

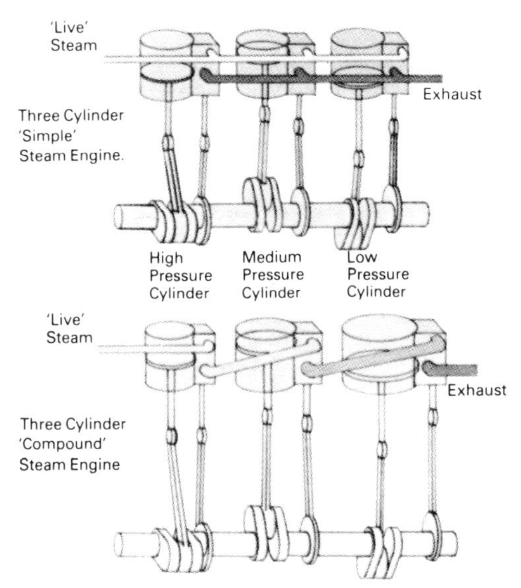

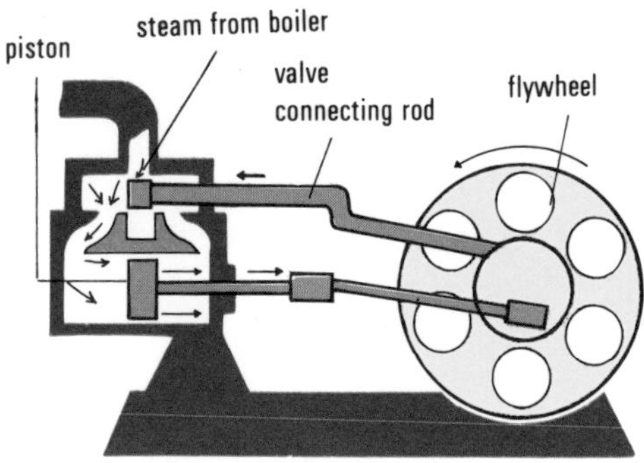

Above: A simple steam engine.

Opposite: The 'Puffing Billy,' which was built in 1813, was used to transport coal from the mine to the dockside.
Below: These three locomotives may look different but they are all steam engines. Engine 'A' is a late 19th century American engine with its own cowcatcher. Engine 'B', the British *Mallard,* set a world speed record of 202kph(126mph) in 1938. Engine 'C' was once used on the Union Pacific Railroad.

pivoted in the middle. As the far end of the beam moved up and down, it operated a pump and lifted water. Newcomen engines were used for a long time to pump water out of mines.

James Watt made great advances in the steam engine, which led to the development of the modern steam locomotive. His engine had a separate cylinder with a piston moving backwards and forwards in it. The piston drove a rod that passed out of the cylinder to drive a wheel, first through gears, then later, with a crank. His cylinder had two holes, or *ports,* in the side, which a sliding valve connected to the steam from the boiler and to the exhaust. This meant that steam was fed first from one side and then the other, so that steam pressure drove the piston in each direction. This was the type of engine that first drove steam locomotives and steamships.

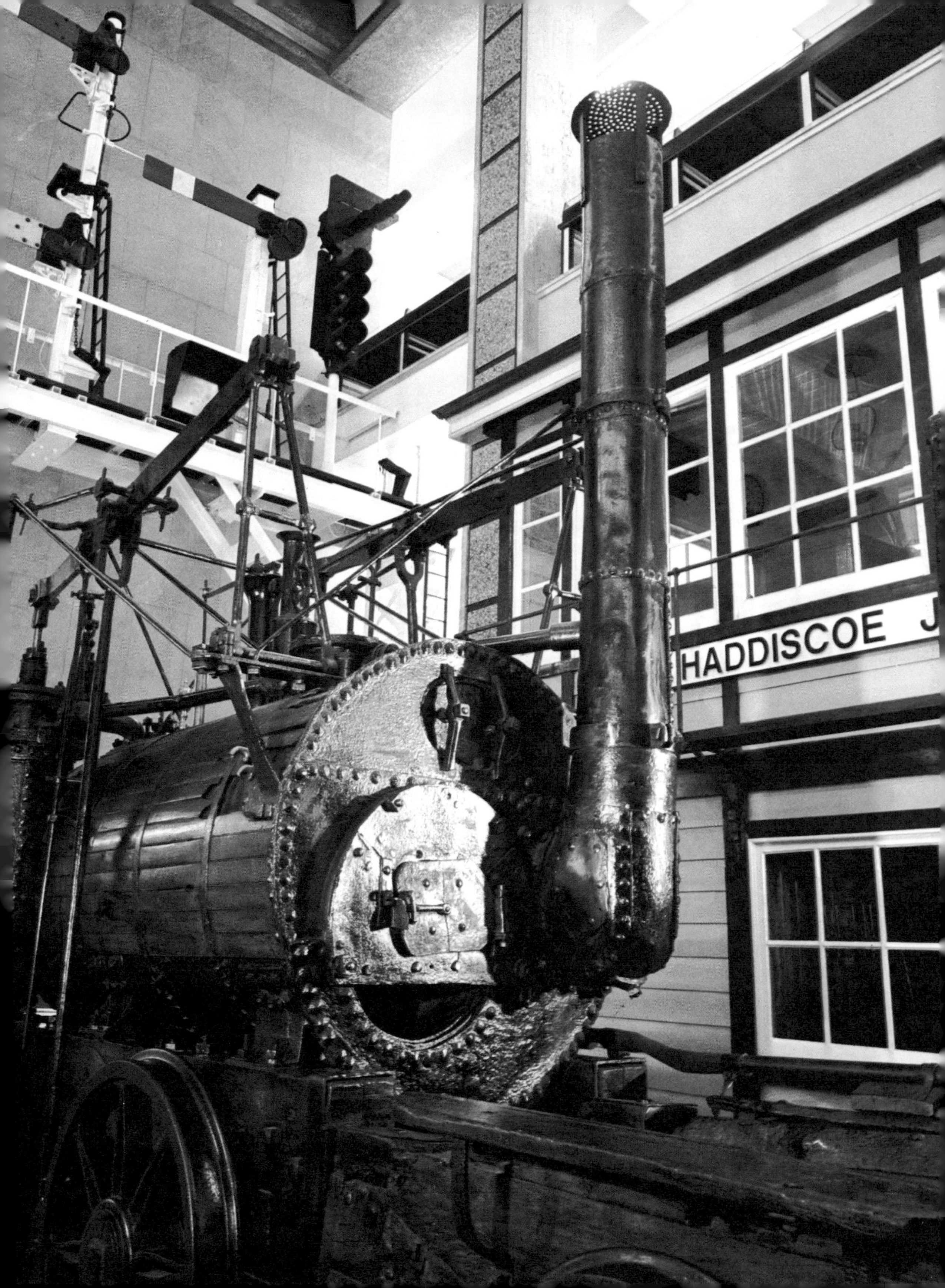
HADDISCOE J

Turbines

Below: The gas turbine engine. (A) accessory drive; (B) compressor; (C) regenerator rotor; (D) variable nozzle unit; (E) power turbine; (F) reduction gear; (G) left regenerator rotor; (H) gas generator turbine; (I) burner; (J) fuel nozzle; (K) igniter; (L) starter-generator; (M) regenerator drive shaft; (N) ignition unit.

Opposite top: This hydroelectric generating system transfers water from a reservoir through a generator turbine into the river which is below.

Opposite bottom: In a water turbine, hydroelectricity is produced as the water inside goes round the volute.

The most efficient means of converting the energy of fossil fuels into electrical energy is to burn them to produce heat which converts water into steam. The force of the expanding steam drives huge turbines which are connected to generators. The fuel may be either coal or oil.

In its simplest form, a turbine consists of a wheel mounted on a shaft. Set around the wheel is a series of buckets or blades. High velocity jets of vapour, steam or heated gas, or moving water from a river or dam, are made to strike the buckets or blades causing the wheel and its shaft to rotate. The turning shaft is then connected to the machine which is to be driven.

For example, it may be connected to a generator to produce electricity or to drive a ship's propeller.

In water-driven turbines, the wheel can be rotated by two different methods. One is called the *impulse method,*

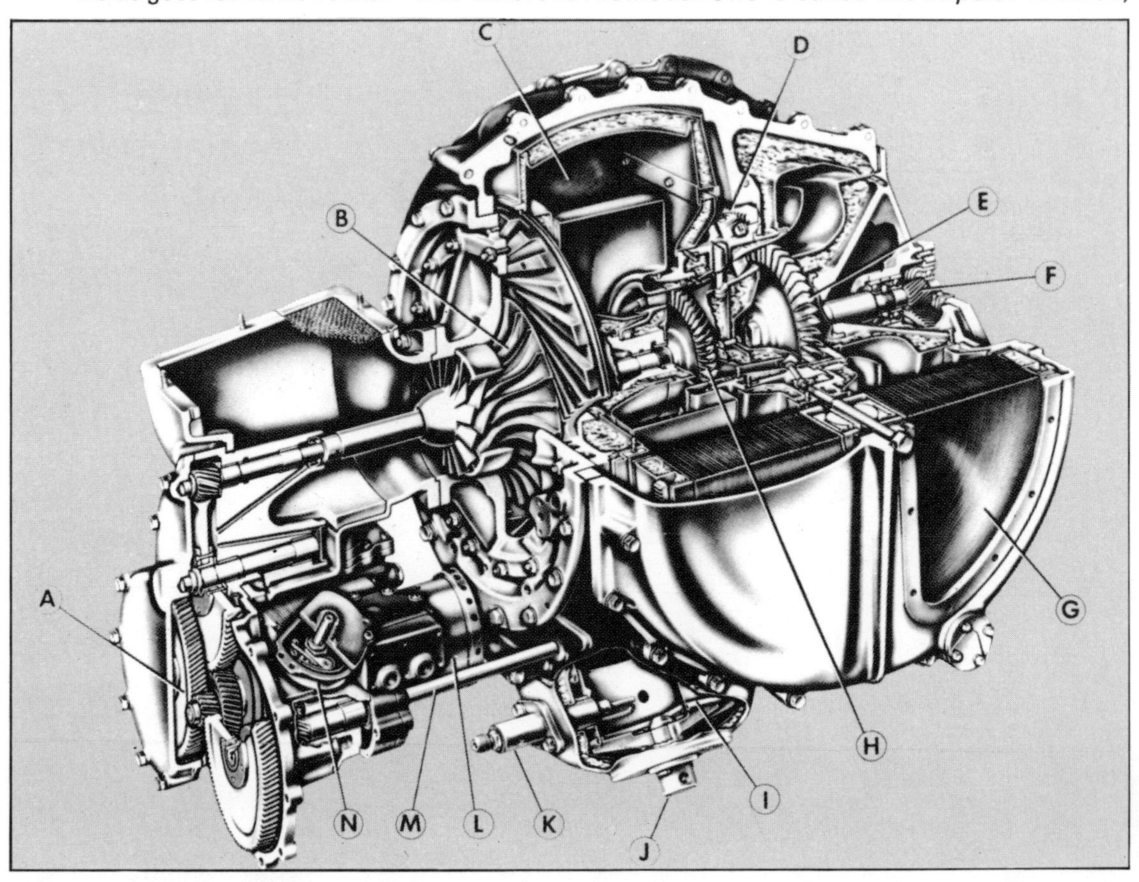

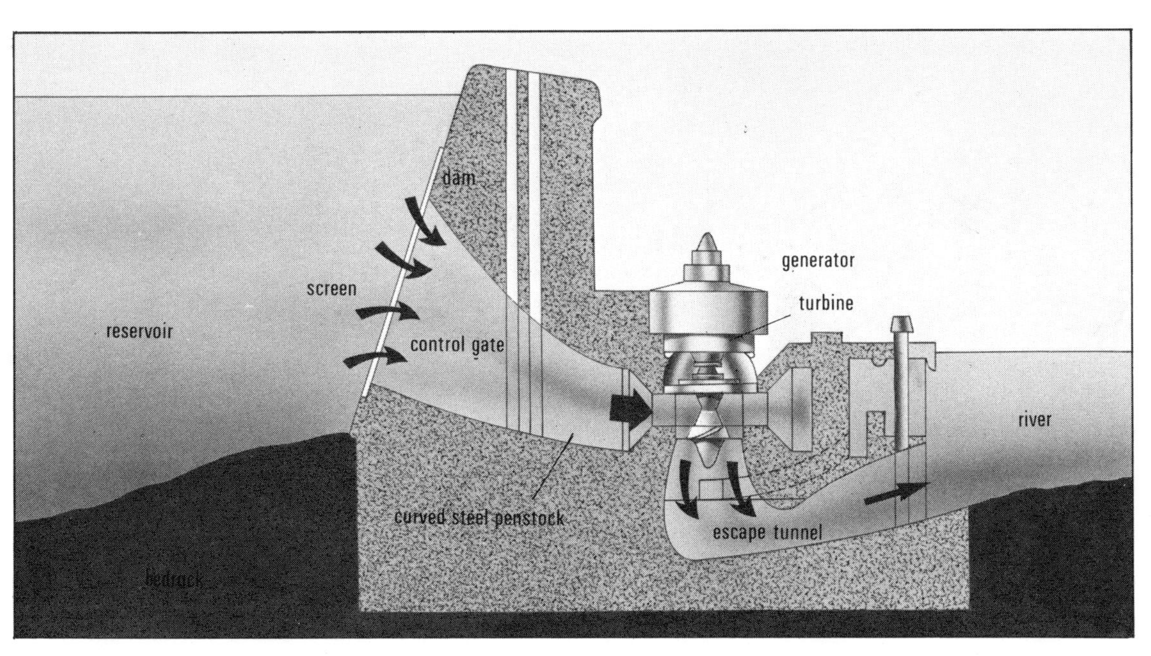
dam
generator
turbine
screen
reservoir
control gate
river
curved steel penstock
escape tunnel
bedrock

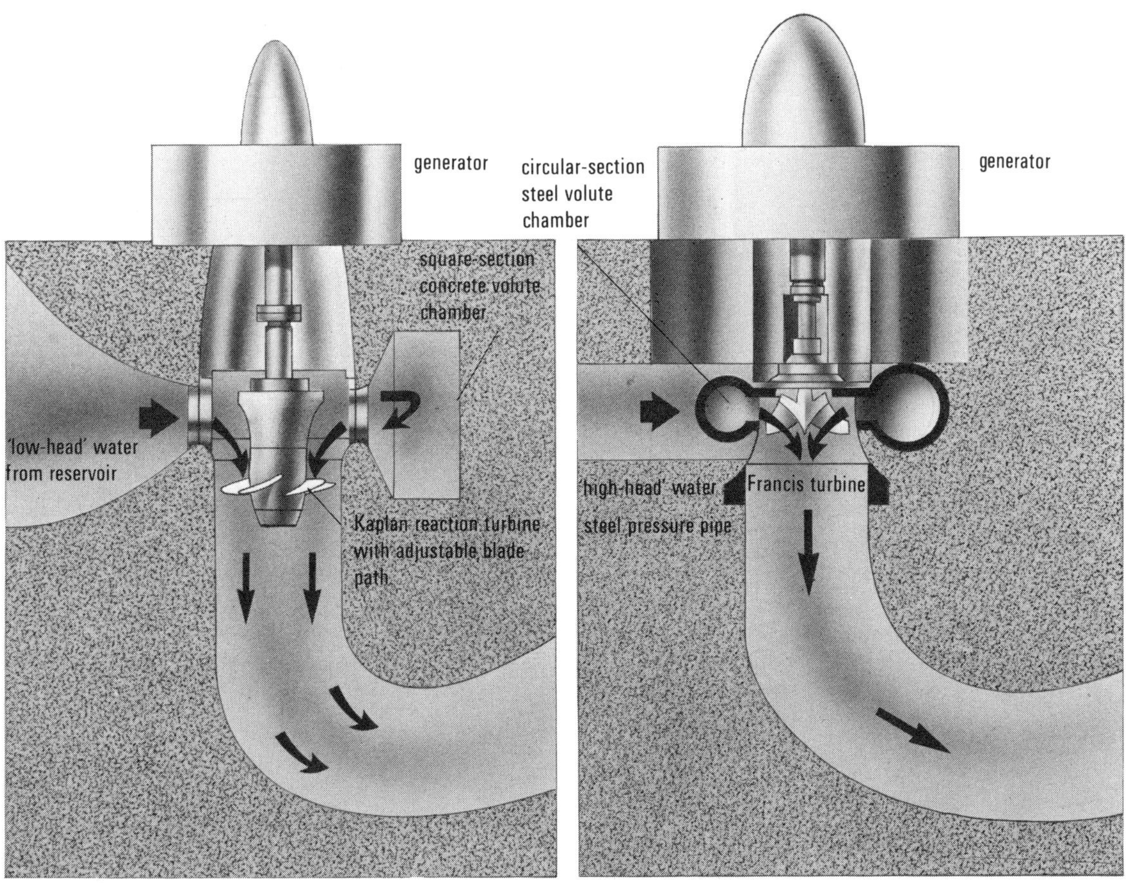
generator
circular-section
steel volute
chamber
generator
square-section
concrete volute
chamber
'low-head' water
from reservoir
Kaplan reaction turbine
with adjustable blade
path
'high-head' water
Francis turbine
steel pressure pipe

Opposite: This space-age room is actually a part of the Western Arm Hydroelectric Power Project in New Zealand. The technician in the foreground is inspecting one of the huge generators.

Below: This engineer, in his protective helmet, works in a nuclear power station. He is inside the low pressure section of a steam turbine.

where the water is forced through a jet and strikes the blades with great force, causing them to move in a direction opposite to the jet stream. The other is called the *reaction method* where the water flows over the blades and has the effect of dragging them.

Steam turbines may be either impulse or reactor types. The steam is produced by heating water in a boiler and then is fed at high pressure into the turbine. Steam turbines can be of the *single-stage* type, that is they have a single wheel, or *multi-stage* with a series of wheels. Multi-stage turbines are more efficient. In a multi-stage turbine each wheel is slightly larger in diameter than the one before it because as it passes through the turbine the steam continues to expand. To help the steam along, a ring of blade-shaped nozzles is fixed to the turbine casing to direct the steam onto the blades of each wheel at exactly the right angle.

Opposite top: Cross-section view of the Rolls Royce turboprop.

Opposite below: A gas turbine. (A) starter-generator; (B) fuel pump; (C) fuel control system; (D) compressor air inlet; (E) regenerator gear-box; (F) combustion chamber; and (G) the exhaust duct.

After leaving the turbine, the steam passes into a condenser which cools it down and converts it back to water. The same water can then be heated up again for another pass through the turbine.

The gas turbine

Gas turbines work in a similar way to steam turbines, but they use the expansion of hot gases instead of steam to provide the energy. The hot gases are produced by combustion of fuel which may be oil, kerosene or natural

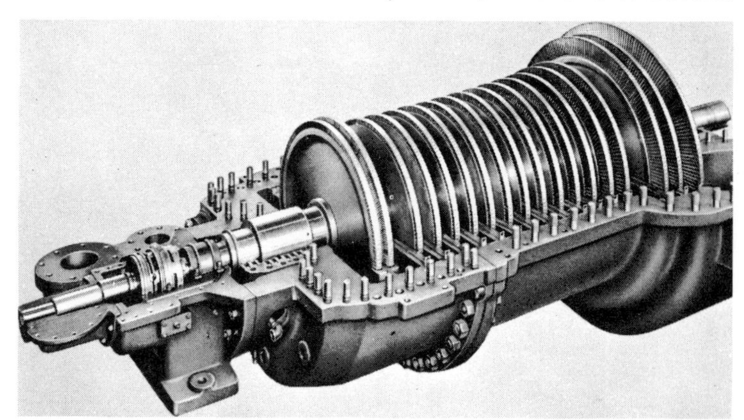

Right: These gas turbines are used for industrial purposes. The single-stage turbine (right) has only one blade, while the multi-stage turbine (above) has many blades. Because it uses more kinetic energy of the gas, it is more efficient than the single-stage turbine.

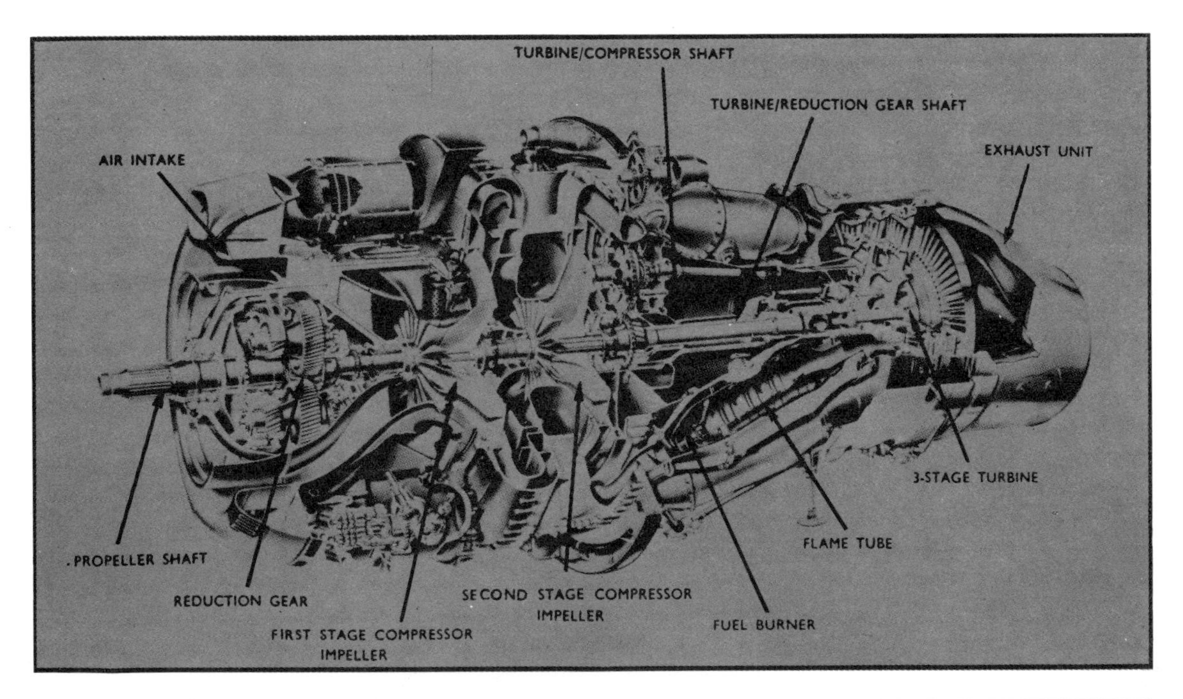

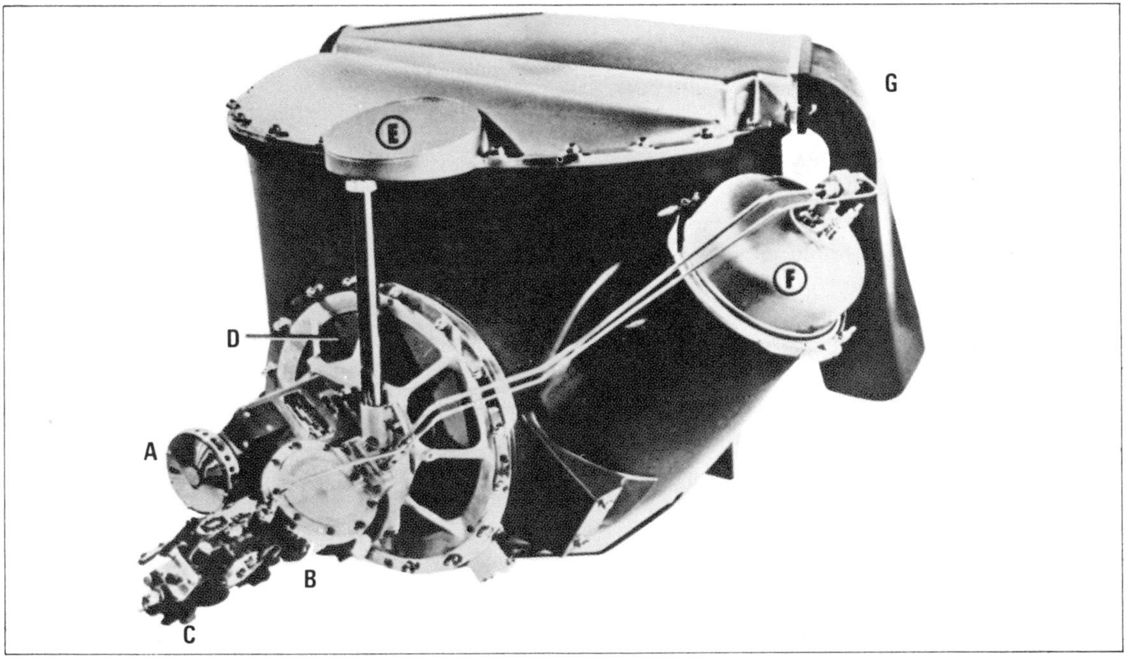

gas. Air is drawn into the front of the turbine, where it passes into a compressor, often in three stages. Between each stage the air is cooled to make the compression more efficient, after which it is heated before it enters the combustion chamber. Here it burns with fuel and expands to drive the turbine

THE ENERGY CYCLE

Opposite: This spectacular picture of a solar prominence was taken by the Skylab astronauts who were orbiting the earth. On the right of the picture, a plume of intensely hot gas is erupting into the corona. By means of X-ray pictures like this, scientists can learn about the sun and the planets.

To perform any action, we need energy. To kick a football, to push a jet plane through the sky, to heat our homes, even to sleep, requires energy. Energy comes to us in various forms. It can be the heat energy stored in fossil fuels which we can use to drive engines or it can be chemical energy supplied by the food we eat. Plants get their energy for growth from the heat and light of the sun.

The sun

In ancient times, many people worshipped the sun, believing it to be the power that gave them life and it is indeed the power of the sun that provides all of our energy. The great reservoirs of oil, natural gas and seams of coal would not have been possible without the warmth of the sun's rays to cause the giant trees and ferns and the tiny sea creatures of long ago to grow on earth. For millions of years wood was the only fuel known to man and he burned wood to cook, to heat, and to drive some of the early steam engines. But wood cannot grow without the aid of the sun.

The food we eat to provide our bodies with nourishment depends on the sunlight for growth. To increase the fertility of the soil so we can grow more food for expanding populations, we add fertilizers. Many of these fertilizers are artificial, being made as a by-product of the petroleum industry. But, as we know, petroleum came from plants which received sunlight. When we eat meat we are again depending on the sun because the animals feed on grass which cannot grow without sunlight.

The process by which plants turn the light of the sun into energy is called *photosynthesis.* The process by which animals and human beings turn food into energy is *respiration.* Digestion is the breakdown and absorption of food. Respiration is the process by which this stored material is converted to energy.

The sun, which provides us with most of our energy, is a star and, like all stars, is made of gas. Its diameter is 109 times that of earth and its volume is 1.3 million times as great. The gas at the centre of the sun is under great pressure and has a temperature of at least 14 million degrees C. On the surface, the temperature is about

Below: This cross-section of the sun demonstrates how small the earth is in relation to the sun. A prominence is erupting through the thin mass of gas called the corona.

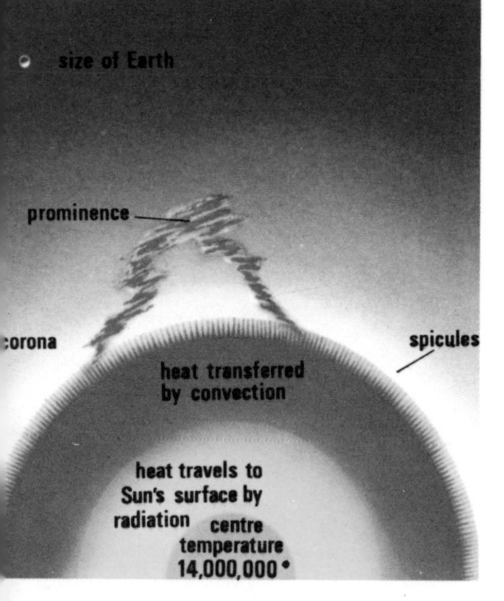

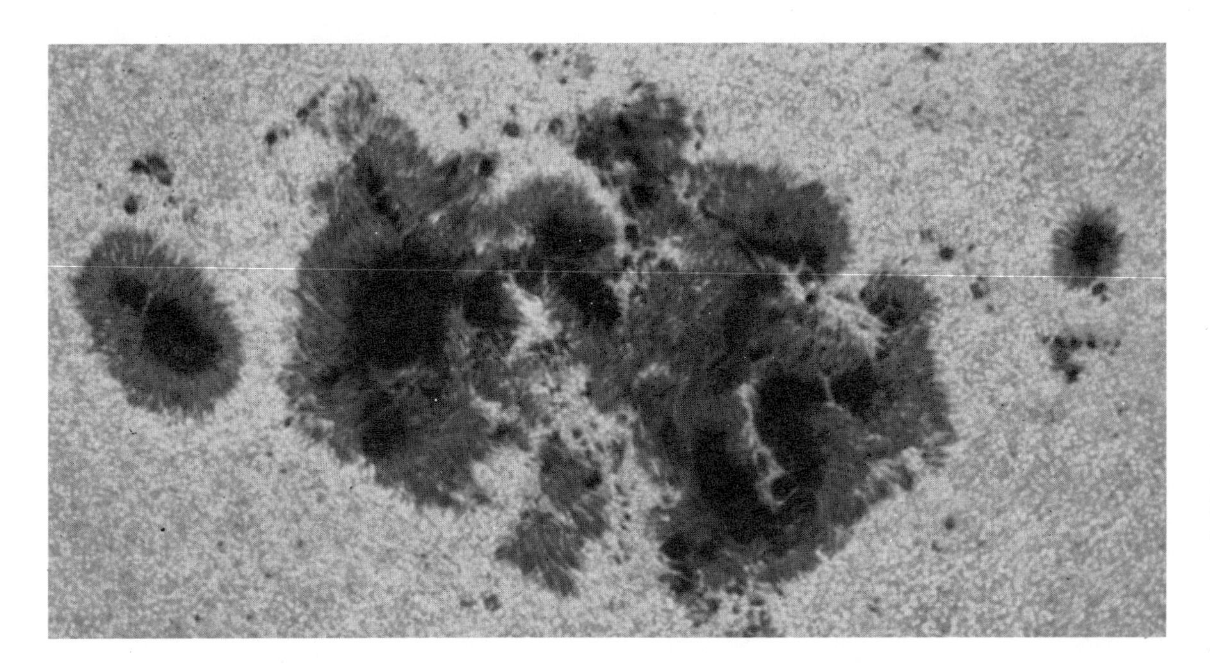

Above: This close-up view of the sun's surface shows clearly that it consists of a network of bright grains with darker, prominent sunspots.

Below: During a total eclipse of the sun, prominences are visible to the human eye. This photograph, taken in Brazil in 1919, is of the famous Anteater Prominence.

5,500°C. By examining the sun through instruments, astronomers have been able to identify the substances that form it. The most plentiful gases present are hydrogen and helium. Also present in the sun are oxygen, nitrogen, carbon, iron and magnesium.

How the sun produces energy

The presence of enormous quantities of hydrogen and helium in the sun gives us a clue to the form of energy production that takes place in the sun and other stars. The process is called *fusion,* and hydrogen is transformed into helium. This transformation takes place at extremely high temperatures and during the process a small amount of mass is 'lost'. This lost mass is converted into energy which is released in every part of the electromagnetic spectrum, from X-rays to visible light rays and radio waves. Of these waves, those that reach the earth give us sunshine, warmth and nourishment for plants to grow, and solar power.

This fusion process is similar to the process that takes place in a hydrogen bomb, but it is continuous and, because of the distance of the earth from the sun and the protection given to the earth by its atmosphere, it is beneficial, not destructive. Without this process on the sun there would be no energy cycle and therefore no life on earth.

INDEX

Page numbers in italics refer to a diagram on that page.
Bold type refers to a heading or sub-heading.